The Development of 8 & 10 cw

MORRIS

LIGHT VANS

1924–1934

by
PETER J. SEYMOUR

with 70 & 105 cu ft
ROYAL MAIL VANS

P&B
Publishing

Contents

	Glossary of terms & definition of abbreviations	3
	Chronology	4
	Preface	5
1	The Early Development of Road Transport	6
2	The Development of the Light Van	10
3	Introduction of Morris 8 cwt Light Vans	13
4	W. R. Morris and the Development of Morris Motors Ltd., Cowley, Oxford until 1936	18
5	Hotchkiss et Cie and Morris Engines Ltd., Coventry	34
6	Osberton Radiators Ltd.	42
	Morris 8/10 cwt Van Radiators – 1924 to 1934	46
	Morris 8/10 cwt Van Radiator Badges – 1924 to 1934	48
	Morris 8/10 cwt Van Radiator Temperature Gauges – 1924 to 1934	49
7	Hollick & Pratt Ltd., Coventry	50
8	The S.U. Company, Ltd. – 1919 to 1936	54
9	Specifications & diagrams, Morris 8 & 10 cwt Light Vans by season 1924 to 1934	59
	Car and frame numbering and identification plates	60
	1924/25/26 season Morris 8 cwt 'Snubnose' Vans	64
	1927 season Morris 8 & 10 cwt 'Snubnose' Light Vans	70
	1928 season Morris 10 cwt 'Snubnose' Light Vans	74
	1929 & 1930 season Morris 10 cwt 'Snubnose' Light Vans	76
	1931 season Morris 8 cwt 'Flatnose' Light Vans	80
	1932 season Morris 8 cwt 'Flatnose' Light Vans	84
	1933 season Morris 8 cwt 'Flatnose' Light Vans	88
	1934 season Morris 8/10 cwt Light Vans	92
10	E. G. Wrigley & Co. Ltd. and Morris Commercial Cars Ltd., Birmingham	96
11	Royal Mail Vans	116
	GWK Royal Mail Vans	126
	Trojan Royal Mail Vans	128
	Ford Model 'T' Royal Mail Vans	130
	Morris and Morris Commercial Royal Mail Vans	132
12	Relationships Between the GPO, Morris Motors Ltd. and their Dealers	150
13	Car to Commercial Conversions	154
14	Morris Cowley Commercial Travellers Car	164
15	Conclusion	166
	Index & Acknowledgements	172

Glossary of terms & definitions of abbreviations

Bullnose — A term now in universal use to describe pre-1927 season Morris Cowley and Oxford cars. The description was never in general use during the 1920s or early 1930s.

Flatnose — A name used to describe 4 cylinder Morris Cowleys & Oxfords (cars) made from the beginning of the 1927 season until the end of the 1931 season and Light Vans built during the 1931, 1932 & 1933 seasons.

Snubnose — A name applied to Morris Light Vans made from the beginning of the 1924 season until the end of the 1930 season.

b.h.p — brake horse power.

h.p. RAC rating — horse power. Calculated by the following formula, using the assumptions that:

1. The piston speed is 1000 ft./min.
2. Mean effective pressure is 90 lb/sq"
3. Mechanical efficiency is 75%

$$\frac{0.75 \times 90 \times D^2 \times 1000 \ N}{168067} = \frac{D^2 \times N}{2.5}$$

Where – N = no. of cylinders
D = diameter of cylinder in inches

The resultant fraction of less than 0.1 of a horsepower shall be omitted.

f.w.b. — four wheel brakes.

Hotchkiss/ Morris engines — 4-cylinder engines made at Gosford St., Coventry and derived from the Continental Red Seal type 'U' engine.

Season/ Model year — The season or model year for Morris vehicles usually commenced at the end of July or during August. This enabled stocks of new models to be built up prior to their announcement in October/ November.

n/a — not available.

GPO — General Post Office.

pt no. — part number.

WD — War Department.

Author's Note

William Richard Morris (1877–1963) became Sir William Morris in 1929 and then Lord Nuffield in 1934. Because Lord Nuffield had not been honoured with a title for most of the period covered by this book, he has been referred to throughout as W. R. Morris.

Chronology 1912–1935

Oct. 1912	W. R. M. Motors Ltd. is formed. (W. R. M. being the initials of William Richard Morris). The 'White & Poppe' Morris Oxford is announced.
April 1915	The first 'Bullnose' Morris Cowley, which was fitted with an American 'Continental' engine, is assembled.
Sept. 1915	McKenna duties introduced. A $33^1/_3$% import duty imposed on cars, but commercial vehicles exempt.
Early 1919	Hotchkiss et Cie of Coventry agree to manufacture engines and gearboxes for Morris.
July 1919	W. R. M. Motors Ltd. is dissolved and Morris Motors Ltd. is formed.
Late 1919	The first 'Bullnose' Morris Cowley with a 'Hotchkiss' engine is assembled.
Jan. 1923	W. R. Morris purchases Osberton Radiators Ltd. and Hollick & Pratt Ltd.
May 1923	W. R. Morris purchases the Coventry branch of Hotchkiss et Cie and forms Morris Engines Ltd.
Jun. 1923	The first 'Snubnose' Morris 8 cwt van, derived from the 'Bullnose' Morris Cowley, is manufactured.
Nov. 1923	'Snubnose' Morris 8 cwt vans are announced.
Jan. 1924	W. R. Morris buys E. G. Wrigley & Co. Ltd. of Soho, Birmingham and forms Morris Commercial Cars Ltd.
April 1924	The GPO purchase their first Morris Royal Mail Van (registered XR 1596): a 'Snubnose' 8 cwt chassis with a 105 cu ft body.
May 1924	The Morris-Commercial 'T' Type, 'Tonner' is introduced.
Mar 1925	The GPO purchase their first Morris-Commercial Royal Mail Van (registered XX 1257): a 'T' Type 'Tonner' with a 240 cu ft body.
Sept. 1925	The Morris-Commercial 12 cwt 'L' Type is introduced.
May 1926	Import duties of $33^1/_3$% are levied on commercial vehicles, in addition to cars, imported from the USA.
Jun. 1926	Morris Motors (1926) Ltd. is formed.
Aug. 1926	The production of 'Flatnose' Morris Cowleys and Oxfords commences.
Oct. 1926	The first 'Snubnose' Morris 8/10 cwt van, derived from the 'Flatnose' Morris Cowley, is manufactured.
Dec. 1926	W. R. Morris purchases The S.U. Company, Ltd.
Feb. 1927	W. R. Morris purchases Wolseley Motors Ltd.
End 1927	The Model 'T' Ford is replaced by the Model 'A'.
Dec. 1927	The first Morris-Commercial 'L' Type 105 cu ft Royal Mail Van (registered YU 2525) is delivered to the GPO.
April 1928	Taxation on fuel is reintroduced, which raises the price of petrol by 31%. Commercial vehicle operators are able to recover half the duty.
Nov. 1929	Production of Morris-Commercial vehicles commences at Adderley Park, Birmingham. This factory had been occupied by Wolseley Motors Ltd.
July 1930	The 'Snubnose' radiator is replaced by the 'Flatnose' radiator on Morris 8 cwt vans.
Oct. 1930	The Morris-Commercial 'T2' 'Tonner', 10 cwt and 15 cwt 'L2's are introduced.
1932	Vehicle production at Morris Commercial's Soho factory ceases.
Sept. 1933	The 'Flatnose' Morris van is replaced by an 8/10 cwt van derived from the 1933 season Morris Cowley, but 'Flatnose' chassis continue to be made specially for the GPO until mid-1934.
Feb. 1935	The first Morris-Commercial 'L2/8' 70 cu ft Royal Mail van (registered BUU 214) is delivered to the GPO.

Preface

For the more than seventy years that the Morris name adorned the motor vehicle, the light van was a pivotal part of its range. Until recently, the Morris van was a common sight in the High Street, whether it be carrying the mail or delivering local goods.

The success of the Morris van was as much down to William Morris himself as it was to the quality of the product. By his own contention, Morris was not an engineer in the formal sense. He did, however, possess great mechanical sense, demonstrated from an early age by his skill in repairing bicycles, motorcycles and motor cars as well as in his designs for his first car, the 'Bullnose' Morris Oxford. He had a firm idea of what he wanted and how his motor cars should be, his shrewd business knowledge ensuring that everything was at the right price.

Part of this legacy was reflected in his choice of suppliers. William Morris was not of the school that believed in manufacturing everything oneself. Rather he preferred to court the manufacturers of the component parts whilst demanding a high standard of construction and good value for money.

It was on that basis that the light vans were designed. Derived from the existing car range, they used those tried and tested components and designs, producing high quality and good value vehicles. Morris knew his market well. He realised that British commercial vehicle manufacturers had largely concentrated on the heavier lorries and that the horse and cart was still the favoured transport for local deliveries. He saw that there was an opportunity in the market for a smaller van. His main rival was Ford, from whom he had learned much about mass production and who had been aided by the freer tax duties on commercial vehicles. The 1920s brought William Morris two advantages; the tax laws on imports were tightened and the pneumatic tyre had become a much more reliable pro-duct. Morris capitalised on these and at the same time appealed to the 'buy British' patriotism of the British goods deliverer. Morris Commercial even boasted its 'T' type truck as having 'a body for every trade and purpose' in an effort to attract many different professions. The reliable and readily-maintained Morris van easily wooed customers such as the Royal Mail, who were to become loyal clients of Morris. Despite their propensity for the standard car, Morris even produced models for the Post Office with equipment special to them!

Peter's book explores not only the range and technicalities of Morris vans over the years but delves into the background of the Morris company, following William Morris's business logic and its relationship with his suppliers, his customers and his rivals. It describes what made the Morris van so popular for light deliveries and such a High Street success for over sixty years.

Stephen Laing
Curator, Heritage Motor Centre, Gaydon

1

The Early Development of Road Transport

Prior to the First World War railways carried most of the goods traffic, with some being conveyed by coastal vessels and some in canal barges, whilst horse cartage was used on the shorter journeys. Only a few traders used motor vehicles for local deliveries but this situation was about to change dramatically.

During the First World War, the abilities and potential of motor vehicles were thoroughly demonstrated when large numbers were used to transport equipment and troops over long distances. Consequently, after the War haulage of goods by motor vehicle developed quickly and its growth was maintained throughout the boom period that followed the Armistice. The new industry also gained a stimulus during a railway strike in 1919.

Provision of vehicles and drivers to permit the new road haulage industry to develop was provided by the sale of surplus army vehicles and the availability of ex-military personnel, who were searching for employment having been trained in the art of driving during the War.

Another factor which helped to establish the road haulage industry was the disorganised state of the railways after the First World War,

Left: A commercial version of the 1914 'White & Poppe' Morris Oxford. The truck body could be fitted with a canvas tilt (as shown on pages 2 and 23) and the vehicle was known as 'Delivery Van Type 2'

This Thornycroft J type was typical of many ex-WD vehicles which became the backbone of many transport fleets during the 1920s. These chassis often worked with a variety of bodies, some even doubling up as weekend charabancs in the cut-throat conditions of the time

coupled with the congested rail network owing to the shortage of rolling stock and high freight charges which had been imposed as a result of the disturbance of the war years. However, when the trade boom collapsed towards the end of 1920, competition within the road haulage industry became intense and rates were often cut to an uneconomic level, with the result that many of the new business ventures failed. Nevertheless, by the mid-1920s trade had improved and the businesses that survived were able to operate more economically, although easy profits had gone, and the road haulage industry started to expand once again.

By then the transportation of goods by rail and road had fallen into three main catagories:

1 With their improved efficiency, coupled with lower freight rates, the railways once again were unrivalled for long haul traffic and the carriage of heavy/bulk commodities. Moreover, rail transport was often much faster than road transport as the solid tyred, rear wheel-braked lorries were restricted to low speeds by law.
2 Motor vehicles were found to be most suitable for the transportation of goods for distances of about 3 to 70 miles, particularly where the traffic was suited to carriage by vans or where a return load could be found for lorries.
3 For short distances of up to three miles, horse haulage was still considered the most economical,

especially in situations where the streets were narrow and congested, the traffic too light to operate a motor van economically, or the road surface too poor.*

As road haulage developed, its great advantage of flexibility and its ability to transport goods from door to door became more and more apparent. Canals and railways could only provide facilities at specific points along their routes. At the end of the First World War, an estimated 41,000 commercial vehicles were in use. By 1935 this figure had risen to nearly $\frac{1}{2}$ million. (*See graph opposite*).

Having seen the growth of road haulage after the First World War, W. R. Morris of Morris Motors Ltd., was prompted to enter into the large scale manufacture of car-derived light vans and purpose-built commercial vehicles, in addition to making cars.

Above: A 1928 season Morris 'Snubnose' 10 cwt Light Van

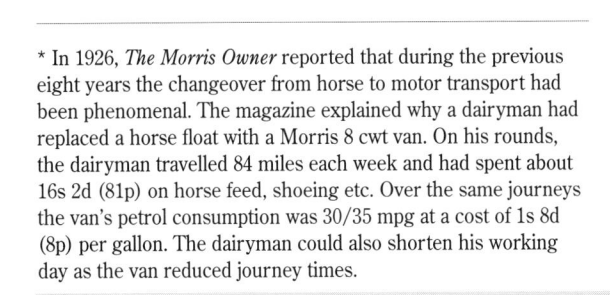

* In 1926, *The Morris Owner* reported that during the previous eight years the changeover from horse to motor transport had been phenomenal. The magazine explained why a dairyman had replaced a horse float with a Morris 8 cwt van. On his rounds, the dairyman travelled 84 miles each week and had spent about 16s 2d (81p) on horse feed, shoeing etc. Over the same journeys the van's petrol consumption was 30/35 mpg at a cost of 1s 8d (8p) per gallon. The dairyman could also shorten his working day as the van reduced journey times.

Commercial vehicles in use in the UK

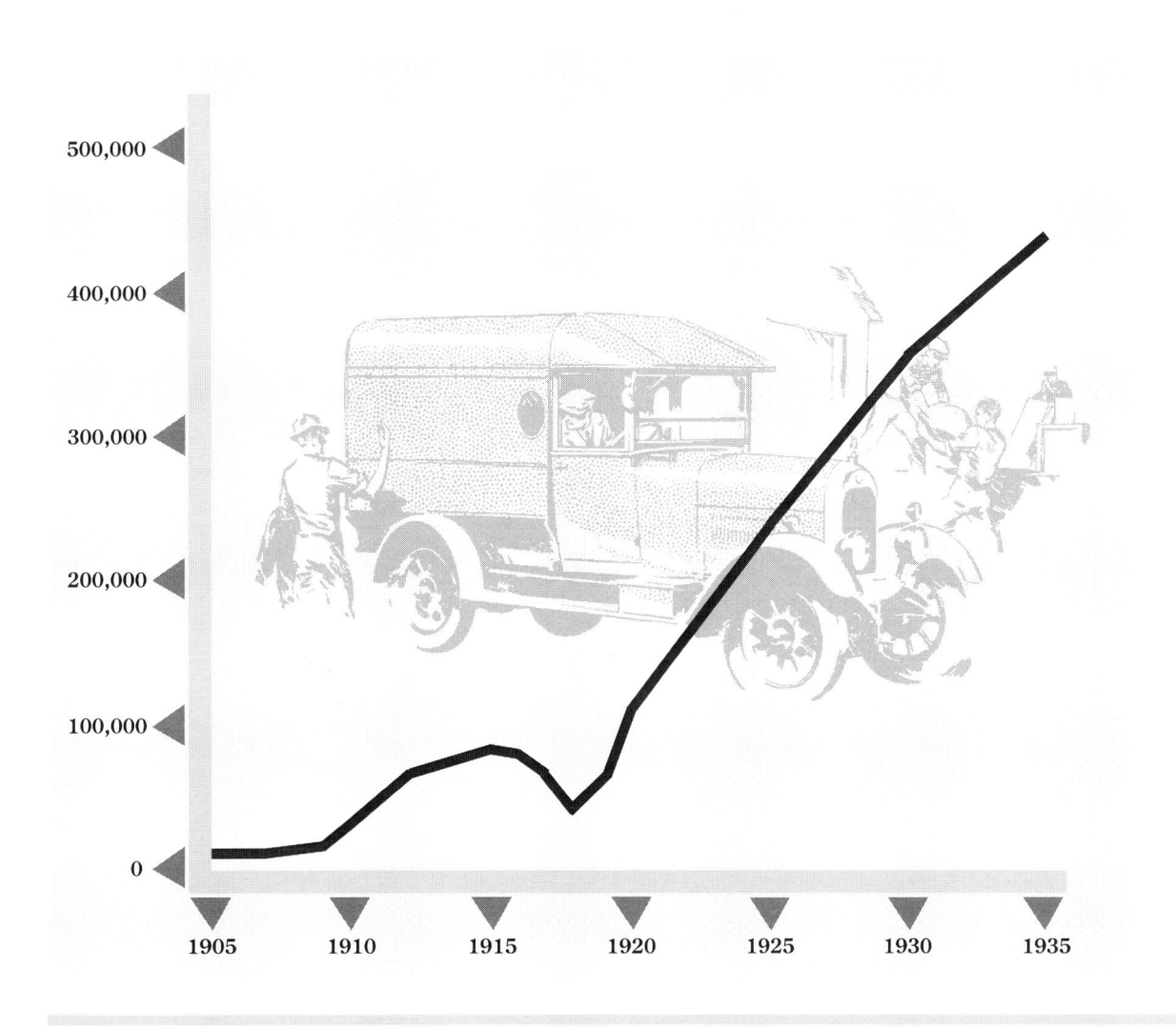

2

The Development of the Light Van

Although they were over two months away from their official announcement, a hint that Morris Motors Ltd., were about to mass produce light vans was given on September 18th 1923, when *The Commercial Motor* published the following article and an illustration of a Morris 8 cwt 'Snubnose' De Luxe Van.*

This article is of particular interest because it not only reflects the situation faced by Morris Motors Ltd. when entering a market dominated by Model 'T' Ford Delivery Vans (not actually named, but apparently referred to) but it also portrays the patriotic sentiments that were evident in Britain at this time.

Is There a Market for the Small British-made Commercial Vehicle?
Interesting Innovations Forecast

'It does not require a great deal of consideration to enable one to arrive at the conclusion that the question "Is a small British van a commercial proposition?" is answered in the same breath as the query "Is it first running costs that are the more important?" Summed up, the imported vehicle usually costs less than the British-built model, but the running costs of the latter are undoubtedly far more attractive to the user.

Signs of Development

Up to the present, although the small touring car of British manufacture has made

---

* Reprinted here with kind permission of *The Commercial Motor*

excellent progress, its counterpart in the commercial field has not become popular. There are signs, however, that show British manufacturers are beginning to cater for the commercial market, and therefore behoves intending purchasers to weigh clearly the pros and cons of the two types. Naturally enough, the question of patriotism has to be dealt with first. From the point of view of the buyer, who, after all, is the ultimate factor, it cannot be argued that patriotism will be the deciding factor alone. It should, however, carry weight. This country gave of its best in the war and it is honourably meeting its debts. Therefore, anything that can be done to help British industries should certainly be done. Take the case of the small tradesman – a typical user of the small van. He must fully realize that, by buying British goods, he is giving work to people who, in turn, will spend their wages with him.

A fact clearly to be grasped is this: Every foreign vehicle sold in this country potentially renders a British workman unemployed for six months. During that period his "dole" comes out of public funds, and, therefore, there is a big argument in favour of buying the home-produced goods. A buyer, however, is not going to pay, say, £20 more for a British van than he would for an imported vehicle of the same potential capabilities merely for patriotism only. Business is too unsentimental an affair for that. What, therefore, does the British van offer in the shape of better value for money than do other types?

In the first place, the average imported vehicle is very much over-engined. It has capabilities for speed that are not wanted; its big, low-efficiency engine has a thirst for fuel, and, although it may easily be said that a driver can be told not to exceed 20 m.p.h., it is very difficult indeed to ensure that he will not do so. One of the chief claims of the British van, therefore, is its economy. The average 12 h.p. four-seater touring car can show a consumption figure of close on 30 m.p.g. when laden and when running in fairly difficult country. There is every possibility that the standard type of van that British manufacturers are concentrating upon will show a very similar figure.

The type of user to whom the light van most appeals is the small tradesman. An analysis of daily work shows that it is seldom that loads greater than 10 cwt. net are carried, and, therefore, the need for the power of the imported so-called "light" van is not only negligible, but, conversely, it means that petrol, oil and tyres are being needlessly consumed every time the vehicle goes out on the road. When a commercial vehicle is in daily use, these factors assume large proportions.

Then as to first cost, it is anticipated that before long there will be produced in considerable numbers 12 h.p. English-made vehicles capable of carrying 10 cwt. at an all-in cost of under £200. Bearing in mind that these vans consist of chassis that have proved their reliability and roadworthiness in every part of the globe, and that the running costs of such vehicles are of the order of 25 m.p.g. of petrol, 1,000 m.p.g. of oil, and 8,000 miles per set of tyres, it will readily be appreciated that, taking the total expenditure over a period of 12 months – that is, adding first cost to 12 months' running – a saving would be effected by buying one of these vehicles in place of an imported type of the most rough-and-ready pattern.

Another big argument in favour of the British van is its longevity. The average American vehicle is well made certainly, but it has not that reserve of wearing powers that enables it to keep on keeping on year after year.

The writer was recently looking over a fleet of small British commercial travellers' vehicles that had been on the road for over two and a half years. They had not been cared for by a headquarters garage, but although, in certain cases, the coachwork required repairing and renovation, the chassis were as sound as could be desired. The British small car chassis has been developed to appeal to the non-

Left: This 1924 season Morris 8 cwt 'Snubnose' De Luxe Van was illustrated in The Commercial Motor *on 18th September 1923, over two months before its official announcement*

❖ 11

technical and unmechanical owner-driver, and, therefore, it has been kept sturdy and simple. Such characteristics are essential in the small van.

During the past year, it is interesting to observe, several private concerns have mounted van bodies on the popular ['Bullnose'] Morris-Oxford and Morris-Cowley chassis, and these have given excellent results. Hewen's* garage at Maidenhead have had vehicles of this type giving yeoman service. The vans look smart, run in a refined manner, and appeal to the prospective buyer as being sound and very serviceable jobs. Further developments in this direction, incidently, may be expected to be made in the near future.

We strongly advise all intending purchases of light vans seriously to consider the claims of the new-type British product, for it presents, in cold terms of value for outlay, a very attractive proposition indeed.'

---

* A Morris main dealer.

Right: The front cover of The Commercial Motor *dated 27th January 1925*

3

Introduction of Morris 8 cwt Light Vans

This delivery van 'Type 1', which was based on the 'White & Poppe' Morris Oxford, had a mahogany-panelled body and was offered in 1914 at a price of £230. The 'Type 1' was the first of many Morris car-derived vans to be manufactured (other views of the 'Type 1' Van are on shown page 23)

Morris Motors Ltd. made their first concerted effort to penetrate the light van market during the 1924 season, when their 11.9 h.p. 'Snubnose' 8 cwt Standard and De-luxe Vans were introduced at the Olympia Motor Transport show, held from 22nd November to 1st December 1923. Prior to this, several 'Bullnose' vans had been manufactured but these were built in limited numbers and probably to individual orders. In fact, car no. 5001, which was the first 'Bullnose' to be erected in 1919 with a Hotchkiss type CA engine, is recorded in the Factory Progress Books as a 'works van'. W.R.M. Motors Ltd. (the predecessor of Morris Motors Ltd.) advertised 'Bullnose' vans in their 1914, 1915 and 1916 sales brochures and these vehicles were derived from the 'White & Poppe' Morris Oxford and the Continental-engined Morris Cowley.

Throughout the 1924 season, only 283 Standard and 171 De-luxe Vans were built, while 753 Standard Vans and 540 De-luxe Vans were produced during the 1925 season, compared with Morris Motors' total vehicle output for those seasons of 27,100 and 48,686 respectively. At this time, therefore, the number of vans being sold was quite insignificant to Morris Motors' overall business and many were in fact bought by their own dealers, who operated them for promotional and publicity purposes. (*See production figures page 17*).

During the early 1920s, the light van market in the U.K. was dominated by

❖ 13

American vehicles such as Chevrolet, Dodge,* Overland, Studebaker and especially Ford, who held much the largest share with their Model 'T' Delivery Van which was being assembled at Trafford Park, near Manchester, from components made in the USA, Canada and Eire, as well as the UK.

Initially, Morris Motors Ltd. found it difficult to penetrate this market as their vans were too expensive. Although the 'McKenna' import duties of $33^{1}/_{3}\%$ were levied on cars, there was no duty on commercial vehicles and this helped Ford to market their Model 'T' 7 cwt Delivery Van in 1924 for £125.0.0, whereas a 1924 season Morris 8 cwt Standard Van was priced at £198.0.0. Also, unlike cars whose road tax was related to their RAC-rated horsepower, the road tax for vans was calculated on their unladen weight, so despite having an engine rated at 22.4 h.p., which was nearly twice that of Morris vans, Ford Model 'T' Delivery Vans were not penalised in this respect.

By 1925, the British government was being pressurised to introduce import duties on commercial vehicles from several captains of industry, including W. R. Morris who had a letter published on the subject in *Motor Transport* of May 18th 1925 as follows:†

---

* The Morris Garages held a Dodge agency until October 1923.
† Reprinted with kind permission of *Motor Transport*.

'Sir,
Viewing the situation from a logical aspect, I can see no reason at all why commercial vehicles should not be included under an extension of the McKenna duties. When these duties were originally introduced [in 1915] it was of national importance that our fighting forces should be able to buy foreign vehicles to use for war transport, because the bulk of British motor factories had then been taken over to produce aircraft engines. That reason for their exclusion from protection has now disappeared, and, since the Armistice, makers of British commercial vehicles have had, if anything, a more difficult task to re-establish their businesses than have even

Right: This advertisement appeared in May 1927

Below: A 1922 Ford Model 'T' 7 cwt Delivery Van

> ## Getting ahead of the foreign truck
>
> **—In economy, efficiency and even price!**
>
> TO-DAY Britain leads in the building of trucks and vans. The great Midland factory of Morris-Commercial Cars has, in a few years, more than answered the challenge of the foreign truck. In economy, efficiency, long life and reliability the all-British Morris-Commercial has scored a decisive victory.
>
> If you keep records which tell you at a glance the comparative running costs of different vans, you will be amazed at the saving of the Morris-Commercial. Its slightly greater first cost rapidly disappears before its wonderful economy in petrol, oil and tyres.
>
> *Morris-Commercial vehicles cater for the following loads:*
> 30-cwt., 25-cwt., 1-ton, 12-cwt., also 30-cwt. and 2-Ton Six-Wheelers.
>
> ### MORRIS-COMMERCIAL
>
> Any Morris-Commercial model can be bought on the most favourable terms. These are explained in our Brochure "Capitalize your Income," which will be gladly sent on request.
> MORRIS-COMMERCIAL CARS LTD., Soho, Birmingham.

touring car makers. They have been faced with the competition of resold war vehicles, and are only just beginning to make progress. It is, of course, important that the nation should have cheap road transport; but this is no reason why British makers should not be given every opportunity to reduce their prices by being able to embark upon increased production, as is possible when they are freed from the menace of foreign competition. Thus, the application of the McKenna duties to commercial vehicles would in the long run attain the end we desire nationally. It would both cheapen transport and at the same time employ more British workmen.

W. R. Morris
Governing Director, Morris Motors Ltd., and Morris Commercial Cars, Ltd.

Inevitably, the price difference between Morris and Ford vans led to the latter being the more popular, but nevertheless, Morris Motors Ltd. persisted with their efforts to gain a respectable share of the light van market and they also promoted their products to large fleet operators such as the Admiralty and the General Post Office. The Morris van supplied to the latter during 1924, for trial purposes, (*see page 132*) eventually led to substantial business being generated for both Morris Motors Ltd. and Morris Commercial Cars Ltd., a company formed in 1924 to manufacture a range of purpose built commercial vehicles, who, like Morris Motors Ltd., also encountered competition from Ford. In August 1924, Ford dealers were able to advertise their 'One Ton Truck' for £132, which was over 40% cheaper than the comparable Morris Commercial 'T' type 'Tonner'.

Three events then took place that dramatically changed the opportunities for Morris Motors Ltd. and Morris Commercial Cars Ltd. *Firstly*, the govern-

> We beg to inform the public that we have been --------
> ## OFFICIALLY APPOINTED
> # FORD SUB-AGENTS
> ### for HARTLAND and District
>
> **FURTHER REDUCED PRICES**
> From August 2nd, 1924
>
> These are the lowest in Ford history and make Ford products the most amazing value ever known in the automobile industry.
>
> Chassis £100
> The New Ford Runabout £120
> Delivery Van £125
> Ton Truck Chassis £107
> One Ton Truck £132
> One Ton Van £137
> Ford Coupé £170
> New Ford "Tudor" Sedan £190
> New Ford "Fordor" Sedan £215
>
> (At Works, Manchester).
>
> All passenger Models are complete with five Demountable Rims and Electric Lighting and Starting Equipment. These models are obtainable without this equipment at £18 less.
>
> *****
> Your Orders Respectfully Solicited
> *****
>
> # HUGGINS BROTHERS
> Hartland

ment decided to introduce import duties on commercial vehicles (33⅓% on those made in the USA), from May 1st 1926 with the result that the number of imported commercial vehicles dropped by 85%, between 1926 and 1927*. *Secondly*, production of the Model 'T' Ford ceased in 1927 and the 7 cwt Delivery Van was replaced by the more expensive 10 cwt Ford Model 'A' van, which was priced at £165.0.0. Morris Motors Ltd. quickly took advantage of this situation by introducing an improved and competitively priced 11.9 h.p., 10 cwt, 'High Top' Van during the 1927 season (*see pages 71 & 72*). *Thirdly*, fuel tax was re-imposed on April 24th 1928, which raised the price of fuel by 31%, although 50% of the duty could be reclaimed by operators of commercial vehicles. This tax especially penalised the Ford Model 'A', with its 24.03 h.p. (3286cc) engine, when compared to the more economical 11.9/13.9 h.p. Morris engines†.

The combination of these events gave Morris Motors Ltd. and Morris Commercial Cars Ltd. a distinct advantage over their main competitors with the result that from 1927, Morris Light Vans began to dominate the market and Morris Commercial Cars Ltd. soon became the largest manufacturer of commercial vehicles in Europe.

10 cwt. Delivery Van, £195.

Above: This Overland advertisement appeared in July 1924

---

* The imposition of import duties on commercial vehicles in 1926 did not have much effect on sales of Trafford Park-built Model 'T' Ford 7 cwt vans and tonners as by then most of the components used in their construction were British made, apart from engine castings which were produced in the Ford foundry at Cork.
† Ford later introduced the 14.9 h.p. (2036cc) Model 'AF'.

Morris 8 cwt Light Van/van chassis & Ford Model 'T' 7 cwt Delivery Van production – 1920 to 1927

Season[8]	1920	1921	1922	1923	1924	1925	1926	1927
Morris 8 cwt Standard Van	–	–	–	n/a	283	753	1269	1872[5]
Morris 8 cwt De-luxe Van	–	–	–	n/a	171	540	552[4]	–
Morris 8 cwt Van chassis & unspecified	–	–	–	n/a	14	255	299	533[6]
Total Morris vans & chassis[7]	–	–	–	24[1]	468	1548	2120[2]	2405

Calendar year	1920	1921	1922	1923	1924	1925	1926	1927
Ford Model 'T' 7 cwt Delivery Van[7]	7167	5229	5019	5029	4172	3198	3468	2285[3]
Combined total – Morris & Ford[7]	7167	5229	5019	5053	4640	4746	5588	4690

1. Production of Morris 8 cwt 'Snubnose' vans commenced at the end of the 1923 season. The first 'Snubnose' van, chassis no. 27412, eng. no. 34782, was erected on 18.6.23.
2. 18 van chassis that were built in June/July 1926 were dismantled in Sept. 1926. These have not been included in the above figures.
3. The Model 'T' Ford 7 cwt Delivery Van ceased production in 1927 and was replaced with the more expensive 10 cwt Ford Model 'A'.
4. Production of the Morris 8 cwt De-Luxe Van ceased at the end of the 1926 season.
5. Of these 1872 vans, 1274 were fitted with the optional four wheel brakes.
6. Of these 533 van chassis, 253 were fitted with the optional four wheel brakes. 18 plain chassis (with two wheel brakes) were converted to four wheel brakes.
7. Note that whilst the number of Morris vans was increasing annually, the number of Fords was decreasing, and that from 1921 to 1927 the combined total remained between 4600 and 5600.
8. See page 3 for definition.

An example of a Ford Model 'TT' One-Ton Truck of which 37,556 were produced at the Trafford Park factory between 1918 and 1927, when it was replaced by the more expensive Model 'AA'.

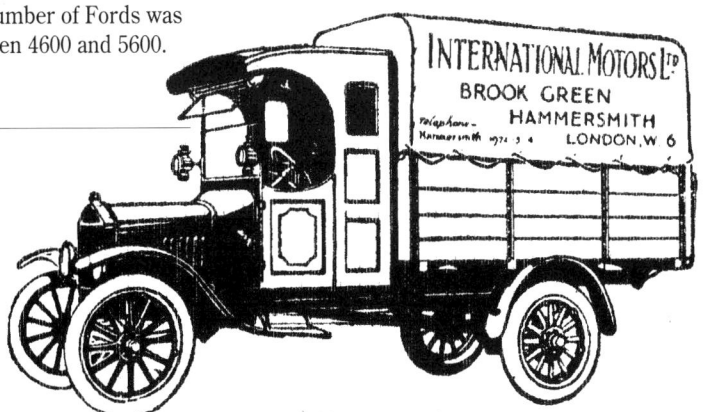

4

W. R. Morris and the Development of Morris Motors Ltd., Cowley, Oxford until 1936

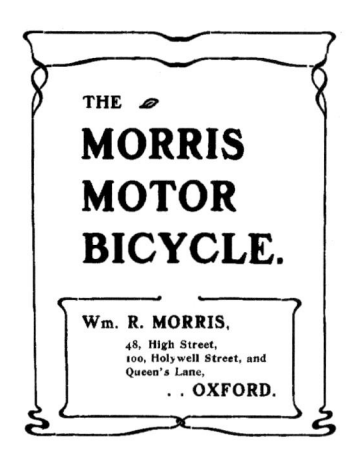

THE
MORRIS
MOTOR
BICYCLE.

Wm. R. MORRIS,
48, High Street,
100, Holywell Street, and
Queen's Lane,
. . OXFORD.

Details of William Richard Morris's early life are well documented in several biographies, and therefore it is only necessary to give a brief summary here. He was born, in modest circumstances, on 10th October 1877 at Worcester, but soon afterwards the family moved to Oxford. William was the eldest of a family of seven children, but four died at an early age. By the 1930s, Morris had become one of the world's richest men but he spent little on himself and preferred to donate most of his fortune to charity. He became Sir William Morris and later Lord Nuffield, taking the name from the small village in Oxfordshire where he then lived.

After leaving school at $15^1/_2$, Morris became apprenticed to a cycle agent called Parker in St Giles, in the centre of Oxford, but following a disagreement over wages, left in 1893. He then started his own business at his father's house in James Street, Cowley, repairing, and later manufacturing, bicycles. Some of these bicycles he raced himself and by 1900 Morris was the holder of seven local championships. Being a self-taught book keeper, his father, Frederick Morris, looked after his son's accounts. The business developed and another premises was soon taken in 48 High Street, Oxford, which was a very good choice for a showroom as it was near to many of his customers in the city. Additional premises were later rented at Queens Lane.

During 1900, Morris bought some castings, machined them himself and built up a 1¾ h.p. engine, which he fitted into a frame of his own making. This machine performed so well that he decided to manufacture and sell motor cycles. However, Morris needed more capital to undertake this enterprise so he entered into a working partnership with a friend, Joseph Cooper, who brought in some much needed additional finance.

The Morris motor cycle

Within a year this partnership faltered and broke up because of a clash of temperament between the two; Morris was always anxious to move ahead and ignore apparent risks, while Cooper remained very cautious, so Morris repaid Cooper his capital. Nevertheless, the two men remained acquainted and Cooper returned some ten years later and was employed by Morris until he retired, by which time he had become foreman of the axle department at Morris Motors Ltd.

Morris soon began to interest himself in motor cars and in 1903, besides advertising himself as the 'Sole maker of the celebrated "Morris cycles and motor cycles"' he could add 'Motor repairs a speciality'. During the same year, he entered into another partnership with a wealthy undergraduate at Oxford University, W. Launcelot Creyke,* and a business-

Above: W. R. Morris, racing cyclist, c.1895 (photo: T. H. Hobbs)

Left: The Morris family taken c.1896 when W. R. Morris (standing behind his father) was aged 19

Right: The premises at 48 High Street, Oxford

*Extracted from The Motoring Annual 1904: CREYKE, W. Launcelot, of Mayfair, London, has owned a 3½ h.p. Renault, an 8 h.p. Progress, two 7 h.p. Panhards, a 12 h.p. Panhard and the famous 'Easter Egg Steamer' of M. Serpollet, which made a record of 75 m.p.h. at Nice in April 1902. Mr. Creyke is a director of the Farman Automobile Co., the Speedwell Co. and is also the proprietor of a motor agency at 16 George Street, Oxford.

❖ 19

man, F. G. Barton*, which traded as 'The Oxford Automobile and Cycle Agency', of 16 George Street, Oxford.

Within months the venture was starting to flounder because the undergraduate, who was officially a sleeping partner, was allegedly spending too freely on entertaining potential customers. The agency was wound up during 1904 and although the revenue generated from the sale of its assets was almost sufficient to pay off its debts, Morris had to settle his share of the trade liabilities amounting to about £50. In addition to losing almost everything that he had built up since starting his own business, Morris had to stand in the rain and bid for his own tools. However, he drew some valuable conclusions from the experience of two failed partnerships, the most far reaching of which was the need to have absolute control of his own business. An illustration of Morris's determination to adhere to this principle occurred 22 years later when he first offered shares (in Morris Motors [1926] Ltd.) to the public. Morris restricted these shares to Preference shares only and, despite the misgivings of his City advisers, he kept all the Ordinary shares himself and, therefore, retained complete control of the company.

Morris resumed trading in 1904 under his own name by taking a small loan from his bank and with help from certain suppliers, who had unshaken confidence in him. Morris continued with the cycle business until 1908 when he sold the enterprise to Mr Edward Armstead*, who not only took over the High Street shop, but also the right to manufacture the motor cycle. This enabled Morris to concentrate on developing his motor car business in a premises at Longwall, known as 'The Oxford Garage', which he had acquired in 1902 to garage cars belonging to wealthy undergraduates. The Oxford directory then listed Morris as a 'motor car engineer and agent, and garage proprietor'.

The enterprise grew rapidly and by 1910 a completely new garage had been built at Longwall. 'The Morris Garage' was now formally adopted as the trad-

Above: 'The Oxford Garage', Longwall. These premises, which W. R. Morris acquired in 1902, were redeveloped in 1910 and the business then became known as 'The Morris Garage'

Below: F. G. Barton

---

* To give himself more time to concentrate on the manufacture of cars at W. R. M. Motors Ltd., W. R. Morris appointed F. G. Barton to the position of General Manager of The Morris Garages in 1913. When Barton left to form the Barton Motor Co. Ltd., the Morris Main Dealer for Plymouth, Edward Armstead was appointed to the position which he held until he resigned in March 1922. Cecil Kimber (*see page 98*), who had been the Sales Manager of The Morris Garages, then took over as General Manager of the company.

ing name of the business. This was changed to 'The Morris Garages (W. R. Morris, Proprietor)' – note that Garages is now in the plural – in 1913 when a further expansion brought showrooms at 36/37 Queen Street, Oxford. Up to 1913, Morris himself had worked in the workshop and his employees were always amused to see how quickly the 'Governor's' head would be under the bonnet whenever a new model of car was driven in.

For some time, W. R. Morris had been thinking about making a car himself. He was therefore anxious to see new designs to add to the knowledge that he had already gained whilst maintaining customers' vehicles, and in this way he developed an understanding of the good and bad aspects of several makes. Also, since starting his own retail garage business, Morris had dealt with customers 'face to face' over the selling, servicing and hiring of motor vehicles and he had acquired an insight into their requirements and expectations.

Consequently, when W. R. Morris decided to expand his business and become a vehicle manufacturer in 1912, he was able to put the experience he had gained at The Morris Garages to good advantage. In addition, having had first hand experience as a motor dealer himself, he was subsequently able to set up a highly successful dealer network for his products, both in the U.K. and overseas. Such understanding of their business gave W. R. Morris common ground when communicating with the staff of a dealership and he was able, therefore, to quickly gain their respect and support.

The prototype of the first Morris car, an 8.9 h.p. (1018cc) 2 seater 'White & Poppe' Morris Oxford, was assembled by W. R. Morris himself with the assistance of some of his employees at The Morris Garage. The vehicle, which *The Autocar* described in its issue of 26th October 1912, consisted of components bought from several outside suppliers; for instance, the engines and gearboxes were made by White & Poppe of Coventry, the 2 seater bodies by Raworths of Oxford and the axles and steering by E. G. Wrigley & Co. Ltd., of Birmingham. The person responsible for the design and manufacture of the components made by Wrigleys and for liasing with Morris, was F. G. Woollard. As a result of this association, E. G. Wrigley & Co. Ltd. was to become Morris Commercial Cars Ltd., and Woollard became the general manager of Morris Engines Ltd., Coventry, as discussed later.

After sufficient orders had been promised, production of the 'White & Poppe' Morris Oxford commenced at Cowley during March 1913 in a disused

Below: When 'The Morris Garage' expanded into these showrooms at 36/37 Queen Street, Oxford in 1913, the business was renamed 'The Morris Garages'

"A Crowning Achievement"

STEWART & ARDERN
MORRIS SALES & SERVICE

1912 — **1937**

Come and see the MORRIS PROGRESS EXHIBITION

THE above interesting photograph, taken after one of the early trials in 1912, shows Lord Nuffield (then Mr. W. R. Morris) at the wheel and Mr. Gordon Stewart on the left of the picture.

We feel that all motorists will be interested to see the collection of cars, representative of the Morris range from 1912 to the present day. This Exhibition is now being held at MORRIS HOUSE, BERKELEY SQUARE, W.1, and has been arranged in order to afford an opportunity to the enormous number of visitors to London during the Coronation period, of seeing this unique display. Admission is free. No official invitation is necessary, and everyone is assured of a cordial welcome.

STEWART & ARDERN LTD
Morris Distributors for the Metropolis & South East Essex
MORRIS HOUSE, BERKELEY SQ., W.1.
AND REGIONAL DEPOTS

GORDON STEWART Chairman & Joint Managing Director   GEORGE H. UPJOHN Joint Managing Director

---

This advertisement appeared in May 1937 and shows W. R. Morris seated at the wheel of a 'White & Poppe' Morris Oxford (registered FC 30) with his accountant, Mr Varney, soon after they had completed a Reliability Trial from London to Edinburgh and back over the weekend of 10th and 11th May 1913 (not 1912 as stated) – note the entry number on the scuttle. In what was its first trial, the Morris Oxford came through without a single adjustment having to be made and W. R. Morris was awarded a gold medal. The picture was taken outside Stewart & Ardern's original premises, at 18 Woodstock Road, London, and shows the joint managing directors of the company, Gordon Stewart and George H. Upjohn (standing on the left and right respectively). Gordon Stewart also held the position of Chairman. The advertisement commemorates the Coronation of King George VI and the 25th Anniversary of Stewart & Ardern's involvement with Morris vehicles.

Of the 1475 'White & Poppe' Morris Oxfords manufactured between 1913 and 1917, only about seven were vans and illustrations of catalogue models are shown here. The 'Type 2' with canvas tilt (shown top right and on page 2) and two views of the 'Type 1' (below)

Military Training College under a new company called W. R. M. Motors Ltd. The first production car was ready at the end of March 1913 and it was collected by Gordon Stewart of Stewart & Ardern Ltd., London.

In September 1915, W.R.M. Motors Ltd. introduced another model, the Morris Cowley, which was available as a 2 seater, a 4 seater, or a van. Initially, most of the mechanical components for the Morris Cowley, including its Continental engine, were imported from the United States and the 'Snubnose' 8 cwt Standard and De Luxe Vans, which were announced in November 1923, were a development of this vehicle.

The early years of the First World War were difficult for Morris. On the outbreak of war, those motor manufacturers that had been established from an engineering business were immediately inundated

❖ 23

with war work but Morris's machining capacity was limited because his factory was laid out for the assembly of bought in components and war work did not come automatically. Consequently, out of necessity rather than lack of patriotism, Morris kept his company going with the trickle of vehicle output that the wartime conditions would allow. At first, Morris could only obtain a relatively small contract for machining bomb cases for Stokes trench howitzers, but this situation changed in 1916.

By a stroke of good fortune, a British minesweeper netted a German mine, complete with its sinker, which was technically far in advance of anything the Allies had at the time. The Admiralty urgently wanted large quantities of sinkers of the same complicated design and this gave the Ministry of Munitions a problem, because all their large engineering plants were already fully engaged on war work. Finally, the authorities accepted Morris's proposal that they should give up the idea of making complete mine sinkers in one place and, instead, use scores of small engineering works up and down the country to make components and then have Morris's staff assemble them into complete units at the Cowley factory. Morris was therefore given a contract in 1916 to make, initially, 250 mine sinkers a week and he became responsible for supplying each sub-contractor with jigs, drawings and standards for the items to be made. The drawings were prepared by Morris's chief designer, Hans Landstad, while Arthur Rowse, who worked for the Ministry of Munitions, arranged for the jigs to be made by firms such as those in Nottingham who normally made machinery for the lace making industry.

Late in 1916, soon after the production of mine sinkers had commenced, the Ministry of Munitions requisitioned all of Morris's factory at Cowley but for the part that was engaged on machining cases for Stokes bombs. W. R. Morris was then officially appointed as 'Controller of Mine Sinker Assembly' with an annual salary of £1,200 (about £48,000 at 1999 values). Although production of Morris Cowleys continued on a small scale during this period, a permit was required from the Ministry of Munitions before one could be purchased.

Morris's production methods proved to be a huge success and his factory eventually achieved a peak output of 2000 mine sinkers a week. The Admiralty had been paying £40 to £50 apiece for mine sinkers, but Morris was

A Continental-engined Morris Cowley carrying a mine and mine sinker, the latter being one of the war products of W. R. M. Motors Ltd. This vehicle has been adapted from the van as illustrated on page 35

Left: Hans Landstad

able to sell them for £21 per unit and for his war work he was given his first public honour; the OBE.

Although W.R.M. Motors Ltd. made a pre-tax profit in 1914 of £13,201, the company showed a loss of £1,107 in 1915. However, the company's performance improved during the three years 1916 to 1918 with average pre-tax profits of £16,883. Morris remained an assembler of modest size and he does not appear to have gained as much from the First World War as some other motor manufacturers. By comparison, the expansion of the Austin Motor Co., Ltd. was such that the company needed to increase its payroll from 2,638 in 1914 to over 21,000 in 1918. Nevertheless, Morris did gain the expertise of Arthur Rowse, a Whitworth Scholar and a brilliant production engineer with a practical bent, who was to play a very important part in Morris's post-war success. Rowse's position during the War with the Ministry of Munitions had brought him into close contact with Morris who had been impressed with his work. After the War, Rowse accepted an invitation to join Morris's organisation as Production Manager and, in 1926, he became a director of Morris Motors (1926) Ltd.

Morris anticipated there would be a large demand for vehicles after the First World War and therefore his main tasks, when peace was declared on 11th November 1918, were to re-organise his factory to cater for this demand and to obtain supplies of components. Consequently, his company showed a loss in 1919 of £7,889, but a pre-tax profit of £41,229 was made in 1920.

Morris and Rowse were able to apply the principles that they had developed for making mine sinkers to making motor vehicles. These new practices, which were a major contribution to Morris's post-war success, enabled the manufacture of component parts to be sub-divided to such an extent that Morris no longer had to buy, for example, a complete rear axle from a proprietary manufacturer such as Wrigleys. This high degree of sub-division enabled Morris to (a) make the car he wanted, albeit a copy of the Continental-engined Morris Cowley (b) produce large quantities of vehicles without massive financial investment and (c) control quality and costs. By the end of 1923, Morris had about 200 sub-contractors, many of whom were machinists, but he retained the responsibility for purchasing and testing raw materials, supplying jigs and drawings and for the quality of the finished product. Although W. R. Morris is generally recognised as an astute businessman, the way he conducted these complicated undertakings also demonstrates his abilities as a skilled negotiator and buyer, who gained the trust and respect of his suppliers.

During the 1920s, Morris bought five important sub-contractors: the engine and gearbox manufacturers, Hotchkiss et Cie (*see Chapter 5*), Osberton Radiators Ltd. (*see Chapter 6*), the body builders, Hollick & Pratt Ltd. (*see Chapter 7*) S.U. Company, Ltd. (*see Chapter 8*) and the axle and steering

Right: A. A. Rowse

gear makers, E. G. Wrigley & Co. Ltd. (*see Chapter 10*). All of these companies became part of his organisation.

The acquisition of these sub-contractors turned W. R. Morris's business from an assembler to a manufacturer of motor vehicles and this situation was reinforced when Morris bought Wolseley Motors Ltd. in 1927.

Morris's policy was that before a new vehicle was released from the factory, it must have been paid for and, because of the credit terms given by sub-contractors, this meant that Morris was usually receiving payment for vehicles before he had to pay for the components from which they had been assembled. Nevertheless, Morris maintained a good relationship with his sub-contractors and he paid them by the due date. He believed that sub-contractors who could rely on prompt payment gave keener quotations.

In effect, this scheme provided the working capital to operate W. R. M. Motors Ltd. but it relied on the ability of their agents to pay for vehicles soon after they were built. When the War ended, Morris realised that it was not enough to rely on just two agents, W. H. M. Burgess and H. W. Cranham, to provide the finance for his expansion plans and so, in July 1919, he placed W. R. M. Motors Ltd. into voluntary liquidation and formed a new company, Morris Motors Ltd., to take over its assets. Morris's prime reason for making the change was to get rid of the restrictive agency contracts with Burgess and Cranham so that he could reorganise his distribution system. Morris believed that this action was necessary not only to sell the number of vehicles that he hoped to produce, but also to provide him with a broader financial base. By 1924, Morris Motors Ltd. had appointed 114 main dealers* who, in their turn, had appointed 400 sub-dealers* and the new distribution system was soon to develop into a worldwide network.

Burgess and Cranham had been natural choices as agents for W. R. M. Motors Ltd. because they were already distributors for 'White & Poppe' engines. In an article about Morris's early history published during 1929, *The Motor* reported that Burgess, whose office was at 40 Glasshouse Street, Piccadilly, London, had obtained orders for 400 'White & Poppe' Morris Oxfords from the blue-prints of the car and had received a deposit of £10 on each vehicle, which he had passed on to W. R. Morris. Most of these orders were probably placed by Gordon Stewart†, of Stewart & Ardern Ltd., London, who was one of Burgess's sub-agents. With the £4,000 supplied by Burgess, plus additional financial backing from the Earl of Macclesfield, Morris was able to commence the manufacture of motor vehicles during 1913. In addition, Burgess is also said to have offered to supply Morris with 'White & Poppe' engines on extended credit. After Morris Motors Ltd. had been formed, W. R. Morris had to pay Burgess compensation as a result of Burgess's action for breach of agreement but the rift between the two men was not permanent and was settled in 1927 soon after W. R. Morris purchased the S.U. Company, Ltd.,

---

* Main dealers and sub-dealers were referred to as distributors and dealers respectively, after the formation of the Nuffield Organisation in 1936.
† Morris Motors Ltd. appointed Gordon Stewart as their main dealer for the London area in 1919.

An aerial view of Morris Motors Ltd., Cowley, Oxford in 1927

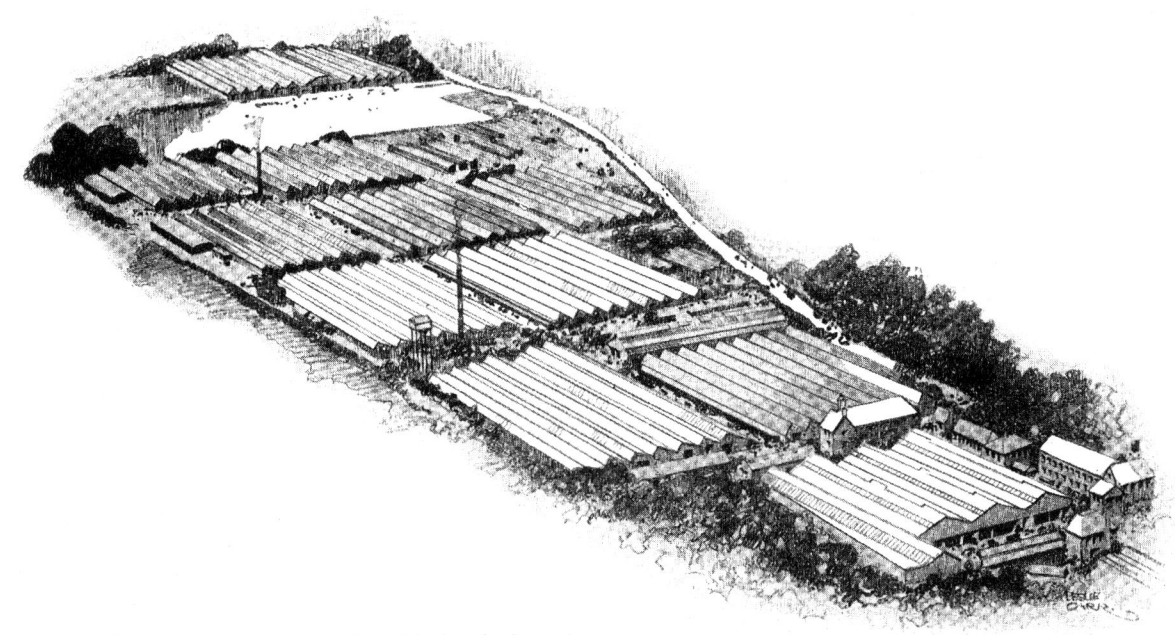

when W. H. M. Burgess became the 'Wholesale Agent' for S.U. carburetters (*see Chapter 8*).

A vital contribution to W. R. Morris's outstanding success, although one which is often overlooked, was made by his financial policies. In an interview for the magazine *System* in 1924, Morris said of his policies: 'They are so simple that you may be inclined to smile at them. Yet I owe very much – more than I can tell – to sticking to them. In the first place, I have insisted on financing the company from the inside. I have never gone to the public for ordinary capital. In consequence all the directors are "still under one hat"'

This policy was in contrast to many other companies. For instance, in 1919 Wolseley Motors Ltd. issued £1.7 million 6.5% debenture stock and by the end of 1921 the Austin Motor Co. had increased their share capital to £3,350,000 from the £650,000 of 1918. In 1920 E. G. Wrigley & Co., Ltd. (*see Chapter 10*) raised their capital from £200,000 to £1.3 million, largely on the strength of promises of orders from Sir William Angus Sanderson & Co. However, when the post-war boom collapsed in 1921, all of these companies were in severe financial trouble because they had saddled themselves with vast debt repayments.

In comparison, the dependance of Morris Motors Ltd. on outside finance to meet its post-war programme was only £29,000 in 7% Preference shares and £20,000 in personal loans. These modest requirements enabled Morris to hold on to the Ordinary shares and thereby retain total control over his business. He gave his views on the subject when he said: 'Personally, I think that the ease with which money can be raised by public flotation in this country may not have proved an unmixed blessing for home enterprise. It has tended to make absentee management easy by shifting responsibility from the original owner to his board, both of whom may so often become absentees. Yet nothing is more pernicious to sound business.'

The wisdom of Morris's policy to retain total control could be clearly seen in 1921. The number of vehicles being sold by his company had declined sharply during the winter of 1920, as a result of the post-war recession. Consequently, there were so many unsold vehicles at the Cowley factory early in 1921 that they were blocking production. Although he had to get his distributors to accept a temporary reduction in their commission on new vehicles, request some suppliers to accept postponment of payment and also increase his bank borrowings to overcome cash flow difficulties, W. R. Morris's sound financial policies enabled him to respond by reducing the prices of Morris vehicles – some by nearly 20% – during February 1921. As a result, vehicle sales were immediately stimulated, pre-tax profits for Morris Motors Ltd, jumped from £144,902 in 1921 to £241,540 (about £9.5 million at 1999 values) in 1922 and the crisis was soon resolved as the bank overdraft was eliminated and all debts paid. Morris was able to make his decision to cut prices swiftly because he had sole control over his company and did not need agreement from a board of directors.*

Left: W. R. Morris and L. P. Lord (with spectacles), the Governing Director and Managing Director respectively, of Morris Motors Ltd. When this picture was taken in 1933, Morris was aged 56 and Lord was 37. At this time, Morris was said to have been earning £2,000 per day (about £80,000 at 1999 values)

The expansion of Morris Motors Ltd. during the 1920s was due to the popularity of the 'Bullnose' Morris Oxford and Cowley, which were reliable and, above all, good value for money. Their popularity was maintained by successive reductions in price which were usually made annually. About 2,000 'Bullnoses' were made during the 12 months ending November 1920, whereas 48,686 (or nearly 1,000 per week) were built during the 1925 season and, in that year, Morris Motors Ltd. became the market leader with 41% of the UK's Total Industry Volume.

However, annual production of Morris vehicles fell in 1928 for the first time since 1920. Pre-tax profits also declined from 1928 (*see page 33*) and the number of vehicles produced by Morris Motors Ltd. com-

---

* In their issue dated 3rd October 1933, *The Motor* declared: 'Some years ago, at a time when most motor manufacturers felt compelled to increase prices, Sir William Morris announced a substantial reduction – a decision which is today admitted to have been the first step in the creation of an immense business affecting the whole future of the British motor industry.'

Right: W. R. Morris (centre), with L. P. Lord, presenting the 100,000th Morris Eight to Mr. Alfred Allen, assistant organiser for the Allotment Gardens for the Unemployed

E. H. Blake

pared to the Total Industry Volume fell from 37% in 1927 to 20% in 1933.

Morris Motors Ltd. were therefore in crisis, and displayed the typical symptoms of a company which had expanded too fast in one sector of the market and had been caught 'off balance' when demand changed to a different sector, as happened from 1928. This change, to lower horse-power cars, was initially brought about by the effects of heavy taxation on motoring, which penalised the owners of large cars and was exacerbated by the depression following the Wall Street financial crash of 1929. Then, towards the end of 1931, the extremely successful Ford Eight was introduced, which soon captured over half of the 'up to 10hp' sector of the market.

By 1933 W. R. Morris was middle-aged and he had lost some of the drive of a decade earlier. Nevertheless, he realised that drastic action was needed to save his business from further decline, so he asked Leonard P. Lord to join Morris Motors Ltd. at Cowley, as managing director, in order to implement a re-organisation.

Leonard Lord, who was once described as 'one of the greatest living production engineers', was to dominate the British Motor Industry for nearly 30 years until he retired in 1961. His response to Morris's offer was to accept the position provided that he had full management control, otherwise he would prefer to remain as managing director of Wolseley Motors Ltd.* Fortunately, Edgar Blake, who had been W. R. Morris's deputy since 1926, agreed to

---

* Leonard Lord was transferred from Morris Motors Engines Branch, Coventry (*see page 37*) to Wolseley Motors Ltd., after W. R. Morris bought this company from the liquidator in 1927, initially to re-organise their machine shops.

retire so Leonard Lord moved down to Cowley during 1933. At the age of 37, Lord faced a formidable task, but he immediately made his presence felt with his tough, unyielding and sometimes blunt manner.

The re-organisation had five main objectives, as follows:

1. To modernise and overhaul of the product range

Morris Motors Ltd. expanded their product range between 1927 and 1933 from two basic types (i.e. the 11.9 h.p. Morris Cowley and 13.9 h.p. Morris Oxford) with 10 body styles, to 9 basic models with 26 body types, in an attempt to reverse the slump in their market share after 1928. Between 1928 and 1933 the 'up to 10 h.p.' sector was the only one to show steady growth, whereas the '11 to 14 h.p.' sector showed a decline (*see table on page 30*). The introduction of the

Percentage of total UK car sales

RAC h.p. rating	1927	1928	1929	1930	1931	1932	1933	1934	1935	1936	1937	1938
Up to 10	22	26	35	36	42	48	60	57	61	60	59	63
11 to 14	55	47	31	24	22	31	23	25	21	25	26	25
15 and over	22	28	33	40	35	22	18	18	18	15	15	13

8 h.p. Morris Minor on September 1st 1928 was, therefore, an important addition to the Morris range.

To counter the increased competition in the small car market, particularly from Ford, Morris Motors Ltd. introduced the Morris Eight, which replaced the Morris Minor in September 1934, less than 18 months after Leonard Lord had given instructions for its design. It was good value, easy to maintain and spacious enough to make it a proper family car and, unlike its main competitor, the Ford Eight, it had hydraulic brakes. By 1938, 230,000 Morris Eights had been sold; more than any other British car before the Second World War.

In addition, Leonard Lord pruned the model range from 9 in 1933 to 5 in 1937 while the range of body styles was reduced from 26 to 10. By the mid 1930s Morris cars were once again beginning to lead the market for quality and value.

2. To modernise and 'specialise' the production facilities

A £300,000 extension and modernisation plan for the Cowley factory was quickly put into effect. Four mechanised assembly lines were installed, incorporating the most modern techniques available. By 1936, expenditure had risen to £500,000 (about £20 million at 1999 values) and many new buildings had been erected. In addition, some factories 'specialised' in the manufacture of certain components. For instance, all engines (except those built by Morris Commercial Cars Ltd. for heavy lorries) would be now made by Morris Motors Engines Branch at Coventry. Also administrative and other functions were centralised, wherever possible.

3. To minimise W. R. Morris's Estate Duties and Super Tax Liabilites

During October 1932, W. R. Morris celebrated his 55th birthday and both he and his advisors were becoming increasingly worried by the effect of Estate Duties on his personally-owned companies because, at that time, the Duties on large estates were 40%.

Immense wealth can also bring other problems, which in the case of W. R. Morris, included claims for Super Tax. The Inland Revenue claimed that Morris, as dominant shareholder, had used his position to retain most of the profits in his companies in order to avoid the Super Tax, that would have been due if the profits had been distributed and become part of his income.

Super Tax was paid in addition to ordinary Income Tax and it was levied against individuals whose annual income was particularly high – £2,000 p.a. (about £80,000 at 1999 values) as from 1922. Unlike Income Tax, which was levied at a fixed rate, Super Tax rose

Right: This advertisement appeared in September 1934

Specialisation gives these 1935 Morris cars a host of advantages

It is not without pride that we of Morris announce our range of cars for 1935. For these cars are the outcome of a bold new policy—SPECIALISATION. You will see the evidence of this vastly improved manufacturing technique in every Morris model—from the smart new Eight to the imposing Twenty-five. Smoother power... greater comfort... better "handling"... more beautiful lines. Precision manufacture has left its unmistakable mark on every detail of these cars. To you who own and drive a car, Specialisation means lower upkeep costs, greater reliability, more economical motoring than you have ever enjoyed before.

NEVER has a small car been so roomy or so stylish as the New MORRIS Eight

Here, at last, is the car that the small-car motorist has always wanted—room, elegance, performance and economy. Because a car is small there is no reason why it must necessarily lose elegance of line and thus cramp body, style and comfort. Here is a "big car in miniature." We mean by that—it is built on the *same lines* as a big car. Its accommodation is of sensible proportions—yet it retains the performance and economical running you require.

MORRIS
THE CAR WITH THE LOWEST UPKEEP COSTS

BUY BRITISH — AND BE PROUD OF IT

---

progressively with rising income and as companies paid only at the standard Income Tax Rate, there were advantages in leaving money in a company.

Two directions for assessment of Super Tax were made in the case of Morris Motors Ltd., for the financial years 1922/23 and 1927/28, and the appeals against them were heard in the High Court during November 1926 and December 1929. Both appeals were won by Morris Motors Ltd., as they were able to show that any profits that had not been distributed had been used for the maintenance and development of their business.

The first assessment made W. R. Morris more conscious of the need for a sound, tax efficient corporate structure, so in June 1926, Morris Motors Ltd. became a public company and a new company, Morris Motors (1926) Ltd., was formed. At the same time, this company absorbed Osberton Radiators Ltd., Morris Engines Ltd., (previously Hotchkiss et Cie) and Hollick & Pratt Ltd., until then all separately owned by W. R. Morris, and these companies then became known as Morris Motors – Radiators Branch, Engines Branch and Bodies Branch respectively. Also, on July 27th 1927, a holding company was registered, Morris Industries Ltd., to enable W. R. Morris to move funds between his companies without incurring tax liabilities and to acquire the S.U. Company, Ltd.

After the second Super Tax assessment, W. R. Morris became convinced that, by a change in the law or for other reasons, it would eventually be impossible to continue his vigorous policy of keeping back profits if the equity continued to be his personal property. If these profits were then to become liable to Super Tax, Morris Motors' reserves might suffer a very heavy and sudden depletion.

W. R. Morris decided, therefore, to merge some of his remaining personally-owned companies and to offer shares on the London Stock Exchange, when conditions were favourable, as these actions were considered to be the best way of minimising both his Estate Duties and Super Tax liabilities.

4. To merge W. R. Morris's personally-owned companies

The general recovery in trade by 1935, following the depression of the early 1930s, brought better stock market conditions which made it feasible for W. R. Morris to merge his personally-owned companies.

The purchase of these companies by Morris Motors Ltd. in 1935 and 1936 was financed by issuing Ordinary shares to the vendor, Morris Industries Ltd., the holding company for W. R. Morris's personal investments. Since its formation, the Ordinary share capital of Morris Motors (1926) Ltd. had stood at 2,000,000 £1 shares, which were all held initially by W. R. Morris but by 1st July 1935 these shares were distributed as follows:

W. R. Morris/Morris Industries Ltd[1]	1,999,995
Leonard P. Lord[2]	1
Reginald W. Thornton[3]	1
Andrew Walsh[4]	1
Cecil Kimber[5]	1
W. M. W. Thomas[6]	1

1 Morris Industries Ltd. was a holding company for W. R. Morris
2 *See pages 29 & 33*
3 Reginald Thornton was from Thornton & Thornton, of Oxford, who were W. R. Morris's auditors
4 Andrew Walsh was an Oxford solicitor and W. R. Morris's legal advisor
5 *See page 98*
6 *See Profiles (page 115)*

Morris Motors Ltd. held an extraordinary general meeting on 1st July 1935, where it was resolved to increase the capital of the company by creating 269,000 Ordinary shares of £1 each. It then bought Wolseley Motors Ltd for £250,000 and the M.G. Car Co. Ltd. for £19,000.

Then, in October 1936, a further 381,000 Ordinary shares were created and Morris Motors Ltd. bought Morris Commercial Cars Ltd. and Morris Industries Exports Ltd. (later Nuffield Exports Ltd.) for £300,053 and The S.U. Carburetter Co. Ltd. for £50,000.

In this way the total nominal issued Ordinary share capital of Morris Motors Ltd. had been increased to £2,650,000 and shares in the Nuffield Organisation, as the merged companies soon became known, were offered on the London Stock Exchange during October 1936.

The only companies then left in W. R. Morris's personal ownership were The Morris Garages Ltd. and Wolseley Aero Engines Ltd.

Although permission to deal was given for the whole of the Ordinary stock – i.e. 2,650,000, 5s (25p) units – W. R. Morris decided to retain three-quarters himself. The shares were made available at 37s 6d (£1.87) and dealings commenced at 39s (£1.95). Public demand was so strong that after a hectic first day's trading, the shares closed at 41s 10½d (£2.09).

W. R. Morris, who became the chairman of the Nuffield Organisation, then divested the greater part of the Ordinary capital into trusts, benefactions and schemes to assist his employees. A majority of the Ordinary stock, if not retained by W. R. Morris, was then held by several charitable trusts and Morris provided that the voting powers for this stock should

R. W. Thornton

A. Walsh

be exercised by three personal trustees, of which he was one. In this way, the new ownership regime left W. R. Morris in the managerial position and he retained control of the company until he retired from its chairmanship in 1952, when aged 75.

5. To Improve Profitability

Between 1920 and 1925, W. R. Morris saw profits from his companies increase from £50,000 to over £1.5 million owing to increased demand for the 'Bullnose' Morris Cowleys and Oxfords. The drop in profits for 1926 (as shown in the following table) was probably owing to the effects of the General Strike,* the costs involved in developing the 'Flatnose' models and changing the production facilities to suit these new vehicles.

From 1928 until 1931, profits showed a downward trend, owing to the change in market conditions discussed previously, but the dramatic increase in profits from 1933 as the result of Leonard Lord's reorganisation is clearly shown:

Pre-tax profits of the businesses controlled by W. R. Morris

Year	Profit	Year	Profit
1920	£50,000	1929	£1,571,000
1921	£145,000	1930	£1,527,000
1922	£252,000	1931	£751,000
1923	£927,000	1932	£971,000
1924	£870,000	1933	£844,000
1925	£1,556,000	1934	£1,168,000
1926	£1,042,000	1935	£1,442,000
1927	£1,290,000	1936	£2,182,000[1]
1928	£1,595,000		

1 Includes 16 months for Wolseley Motors Ltd. and the M.G. Car Co., Ltd., and 17 months for Morris Industries Exports Ltd.

* Out of a workforce of 3000 at the Cowley factory, only three persons joined the General Strike in 1926.

Leonard Lord's Achievement

By 1936, Leonard Lord had achieved the five main objectives with great success. As a result, Morris Motors Ltd. had been transformed into the largest and technically the most advanced vehicle manufacturer in Europe, with the Cowley factory capable of producing 2000 vehicles a week, which was more than the entire German motor industry. At the same time, there were record levels of pre-tax profit, as already shown, and the value of sales jumped from £11,379,000 in 1933 to £21,124,000 in 1936.

The climax of the reorganisation occurred in 1939 when Morris Motors Ltd. became the first British manufacturer to build 1,000,000 vehicles.

Having achieved these five objectives, Leonard P. Lord resigned his managing directorship of Morris Motors Ltd. in 1936 and went to the USA to study the production practices of several motor manufacturers. On his return, Lord agreed to administer a £7 million trust fund to aid areas of high unemployment for W. R. Morris. In 1937, Lord joined the Austin Motor Co., Ltd., at Longbridge, Birmingham and became Chairman of the company in November 1945. When the British Motor Corporation was formed in April 1952, after the merger of Morris Motors Ltd. and the Austin Motor Co., Ltd., W. R. Morris became the Chairman of the new organisation with Leonard Lord as his Deputy and Managing Director. Six months later, when W. R. Morris (then aged 75) stood down to become honorary president, Leonard Lord (later Sir Leonard and then Lord Lambury) became the Chairman of the British Motor Corporation, a position he held until he retired in 1961. Leonard Lord died in 1967 at the age of 71.

5

Hotchkiss et Cie & Morris Engines Ltd., Coventry

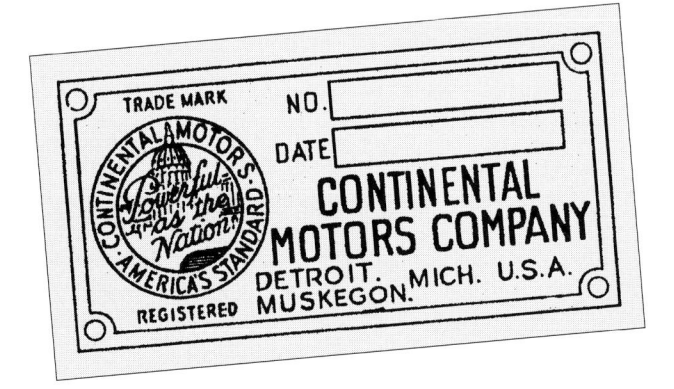

W. A. Frederick

During 1914, W. R. Morris visited the Continental Motor Manufacturing Co. of Detroit, U.S.A. and placed an initial order for their Red Seal type 'U' 1557cc engines. Morris travelled with Hans Landstad, a Norwegian engineer who had taken leave from his employer, White & Poppe of Coventry, to study American production methods. On his return to the UK in 1915, Landstad became Morris's chief designer. At the time of Morris's and Landstad's visit, Continental's chief engineer was Walter A. Frederick who had been responsible for introducing a new generation of 4 and 6 cylinder engines, including the Red Seal type 'U', which was announced in the American motoring press at the beginning of September 1914.

Morris also ordered gearboxes, axles and steering sets from other American companies and despite the freight and insurance charges involved in transporting these items across the Atlantic to Cowley, they were still considerably cheaper than comparable units made in the UK. Although some of the components were lost at sea as a result of enemy action during the First World War, 1,485 Continental engines, together with the American gearboxes, axles and steering sets, were installed into Morris Cowleys, between 1915 and 1919.

A severe setback occurred on 29th September 1915 when Reginald McKenna, the Chancellor of the Exchequer, imposed a 33$\frac{1}{3}$% import duty on so-called

A 1557cc (12.1 h.p.) Continental 'Red Seal' type 'U' engine, coupled to a gearbox made by the Detroit Gear and Machine Co.

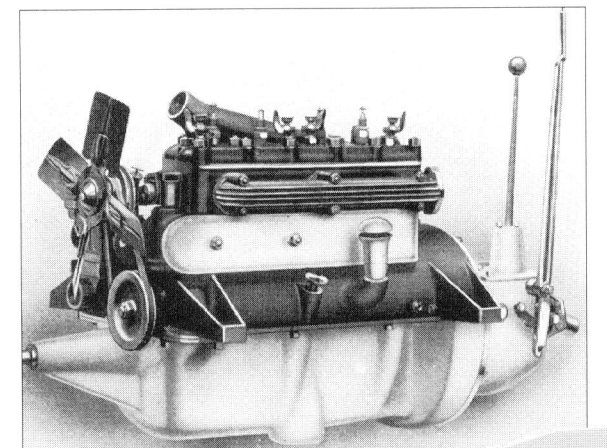

luxuries which included cars and components, although tyres and commercial vehicles were exempt as they were considered essential to the war effort. Then on 27th March 1916 the importation of cars and components from the USA was prohibited unless an import licence had been obtained. These restrictions did not have an immediate effect on Morris's business because in 1916 his company was given a contract to manufacture mine sinkers for the Admiralty (*see page 24*). However, when the War drew to a close Morris began to search for someone to make the

Of the 1,485 Morris Cowleys made between 1915 and 1919 with an American power unit (as shown above) only 37 (2½%) were vans and this is one of them. The 'Snubnose' 8 cwt Vans introduced in November 1923 (see Chapter 3 & page 64) were developed from this vehicle type

American engine and gearbox in the UK, as he had now obtained the rights to manufacture them and a number of companies were approached.

The Hotchkiss company of France, who were the makers of the famous machine gun, had hurriedly transferred production to England during the First World War when it looked as if their St. Denis factory was going to be overrun by the Germans. Consequently, a factory was erected in Gosford Street, Coventry and both machines and key people were brought over to England so that production could start as soon as possible.

At the end of the First World War the factory suddenly became short of work and M. Benet, the Director General of Hotchkiss et Cie and H. M. Ainsworth, the Works Manager at Gosford Street, then made contact with W. R. Morris in Oxford as they had heard that he was looking for an engine supplier. Subsequently, Hotchkiss agreed to copy the Continental engine and Detroit Gear gearbox and to manufacture them for 'under £50'. The relationship created at that time eventually turned into good fortune for Morris.

The first engine to be supplied by Hotchkiss to Morris Motors Ltd. was delivered during the middle part of 1919 and was designated the type 'CA'; the derivation of this is not known but suggestions for the meaning of the initials include 'Continental Type A' and also 'Coventry Type A'.

There were many differences between the type CA engine and the Continental engine and they have few interchangeable parts because they differed dimensionally. Continental worked to an Imperial Standard whereas Hotchkiss/Morris power units were made to Metric measurements with French Metric screw threads. However, the hexagons of the nuts and bolts were to a British Standard.

The sales volume of Morris Cars continued to rise until the early part of 1922 when the supply of engines was only just sufficient to meet the level of car production at Cowley. In the autumn of 1922 Morris asked Hotchkiss to raise engine production to 500–600 per week, but Hotchkiss refused and were unwilling to make more than 300 per week because an expansion in England would have needed capital that they preferred to use in France.

At this time Hotchkiss were also making engines

An ariel view of Morris Engines Ltd., Gosford Street, Coventry in 1923

Left: This 1920 advertisement shows part of the Hotchkiss factory at Gosford Street, Coventry and a type 'CA' engine and gearbox

and gearboxes for other manufacturers, as they did not wish to be dependant on one customer.

This state of affairs no doubt irritated Morris and after a lot of argument a Hotchkiss director asked him 'why don't you buy the works?' which he subsequently did, paying a total of £349,423 (about £14 million at 1999 values) for the plant and premises. In addition, Morris also acquired an experienced workforce amongst whom was Leonard Lord, a man destined to dominate the affairs of Morris Motors Ltd., the Austin Motor Co. Ltd. and the British Motor Corporation until 1961 (*see Chapter 4*). Leonard Lord had joined Hotchkiss in 1919 from the Coventry Ordnance Works and was employed in the drawing office when Morris bought the Gosford Street factory, but he soon became responsible for the design and purchase of machinery.

W. R. Morris, therefore, formed Morris Engines Ltd. (later Morris Motors Ltd. Engines Branch) in January 1923, to take over the Coventry business of Hotchkiss et Cie, and he immediately set about a re-organisation. To carry out this task Morris appointed F. G. Woollard, who was then aged 40, as General Manager. Woollard, who had been working for E. G. Wrigley & Co. Ltd., was already well known to W. R. Morris (*see Chapter 10*) and he was recognised as a brilliant production engineer with advanced ideas. He was a pioneer in flow production since he had been experimenting with machine shop layout for some years. The results that Woollard obtained at Gosford Street were largely due to his expertise in managing man and machines.

W. R. Morris invested £300,000 in extending the Gosford Street factory and in buying new machines, some of which were for machining cylinder blocks and which formed a hand-transfer line. This series of machines carried out 53 operations in turn and were capable of producing a fully-machined cylinder block, including bearing blocks, crankshaft bearings and studs, at the rate of one every 4 minutes. Then in 1924, the first automated transfer machine was installed for the production of gearbox casings.

These machines were something that Morris Engines Ltd. were justifiably proud of and they encouraged the public to see the process with the result that the factory became a model for all British industry. American engineers judged the production methods to be twenty years ahead of their time.

Having re-organised the layout of the works, Woollard introduced a 24 hour working day for a 5 day week and this was achieved by 3 shifts of 8 hours each, being 6am to 2pm, 2pm to 10pm and 10pm to 6am. The shifts were timed to suit public transport as the 750 persons on each shift (of whom about 7% were quality inspectors) were commuting from as far away as Wolverhampton.

The faith that Morris had in the abilities of F. G. Woollard, who gained a reputation for being an inspiring and warm-hearted man with great presence, was soon to become apparent. At the time Morris acquired the Gosford Street factory in May 1923, it was producing about 300 power units per week. By December 1924 a weekly production of 1,200 power units had been achieved.

W. R. Morris's decision to purchase licences to make American engines and gearboxes in the UK and then to buy the Gosford Street factory were fundamental to his future prosperity. This combination assisted him to develop an excellent reputation for his products with the result, as already mentioned, that Morris Motors Ltd. achieved 41% of the UK's 'Total Industry Volume' during 1925 to become the market leader.

F. G. Woollard, M.B.E.

Right: The cylinder block hand transfer line, installed during 1923 at the Gosford Street factory, carried out 53 machining operations in turn. The process took 224 minutes and fully machined blocks were being produced at the rate of one every four minutes

❖ 39

Hotchkiss/Morris 11.9/13.9 h.p. engine types

Type	RAC h.p.	Fitted to
CA	11.9	Morris Cowley & Oxford
CB	11.9	Morris Cowley & Oxford, Morris Light Vans ('Snubnose' & 'Flatnose'), Morris Commercial 'L' type
CD	11.9	Gilchrist Cars, C. Kimber's Morris Cowley Special (FC 7900)
CE & CF	13.9	Morris Oxford, M.G. 14/28 & 14/40, Morris Light Van ('Flatnose'), Morris-Commercial 'T' & 'LT' types.
CG	13.9	Morris Light Van ('Flatnose' – 1931/2 seasons)
CH & CL	11.9	Morris Cowley
CJ & CM	13.9	Morris Cowley
CK	11.9	Morris 70 cu ft Royal Mail Vans ('Flatnose')
CN	13.9	Morris Light Van ('Flatnose' – 1933 season)
CO	13.9	Morris-Commercial G2 Taxi cab
CP	13.9	Morris-Commercial 'T2' & 'L2' (early models)
CQ	13.9	Morris-Commercial 'T2', 'T2F', 'L2', 'L2/8' & DCTM Mk 1 Military
CR	13.9	Morris-Commercial G2 Taxi cab
CS	13.9	Morris 8/10 cwt Van (1934 season)
CSDC	13/9	Morris-Commercial 'T3', 'L3', 'L3/8' & DCTM Mk II Military
IM & MM	11.9/13.9	Morris Industrial & Marine

All these engine types (except IM & MM) share the same serial number sequence. An engine's type and serial number can be found on its identity plate/disc as shown.

A 'Maxfield' tyre pump was mounted on to some type CB & CE engines, installed in Morris-Commercials. The pump engaged with the magneto drive shaft via a dog clutch.

Engine specifications

RAC h.p.	Bore & Stroke	Cubic Capacity	No of cyls	Max BHP @ RPM* Non-turbulent	Turbulent
11.9	69.5 × 102mm	1548cc	4	26 @ 2600	31 @ 3400
13.9	75 × 102mm	1802cc	4	32 @ 3000	36 @ 3400

* 'Non-turbulent' and 'Turbulent' 13.9 h.p. engines were sometimes referred to as '14/28' and '14/32' respectively. 'Non-turbulent' type cylinder heads have their sparking plugs in 'pairs' above the inlet valves, whereas the sparking plugs on the 'Turbulent' types are equally spaced. 'Turbulent' engines, which were introduced in 1930 for cars and light vans, were fitted with 'Ricardo' cylinder heads which required smaller (33mm dia.) valves.

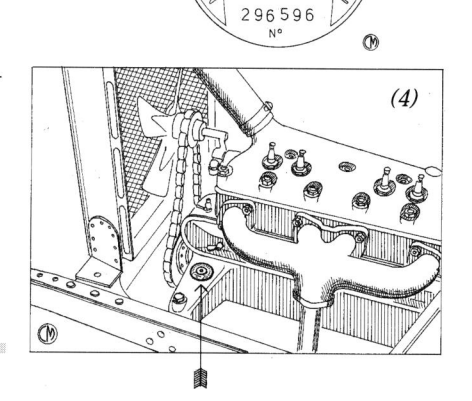

Left: (1) The style of brass engine identity plates fitted to the side cover of 11.9/13.9 h.p. engines until c. May 1923

(2) The style of brass identity plate fitted to the side cover of 11.9/13.9 h.p. engines from c.May 1923 to early 1926

(3) Detail of the brass identity discs fitted to 11.9/13.9 h.p. engines from early 1926. Note that the inner disc shows the engine type and the outer the serial number

(4) The brass engine identity discs, shown (3), were rivetted to the front l.h. engine bearer

Above: A consignment of 11.9/13.9 h.p. engines and gearboxes arriving at the Cowley factory from Morris Engines Ltd., Coventry in 1925

6

Osberton Radiators Ltd., Oxford

When W. R. Morris first started to make motor cars before the First World War, he purchased radiators from several companies including Doherty Motor Components Ltd. of Coventry. This arrangement continued after the war but W. R. Morris soon ran up against the problem of getting the necessary quantity of radiators to suit the output of Morris vehicles. To overcome this difficulty, he assisted H. A. Ryder and A. L. Davis, who had been working foremen at Doherty's in Coventry, to set up a radiator manufacturing operation in Oxford. The business commenced on 5th May 1919 in a small workshop located in Alfred Street, Oxford.

Just over three months later, during August 1919, the business moved to a former roller-skating rink in Osberton Road, located in the Summerton part of Oxford and the business then became known as The Osberton Radiator Co. The roller-skating rink in Osberton Road had been opened on 11th December 1909 but closed three years later. W. R. Morris then acquired the premises and they became a garage for the six Daimler 'buses operated by the Oxford Motor Omnibus Company – a company set up by Morris in 1913. When Morris relinquished his interests in the 'bus operation, the premises were occupied by the Oxford University Flying Squadron, but, by 1919, they had become vacant.

Since The Osberton Radiator Co. was an independent company, commercial common sense dictated that it should not rely solely on business from Morris Motors Ltd. and it therefore took on work from other customers. Despite increasing the number of its staff and radiator output, the company found it difficult to keep pace with the steadily rising demand from

H. A. Ryder

A view of the Osberton Radiators Ltd. factory in Osberton Road, Oxford c.1925. Note the two piles of honeycomb strips in the foreground for making radiator cores. Completed 'Bullnose' cores can be seen being grouped together on a jig for roll dipping in solder. The large presses in the centre of the picture are for making radiator shells and the line shafting on the right is driving smaller presses. The assembly of bonnet panels, which were pressed at the factory, can also be seen

❖ 43

Morris Motors Ltd. at Cowley, so, in order to overcome this situation, W. R. Morris took complete control of The Osberton Radiator Co., for which he paid £15,000 (about £600,000 at 1999 values) on 1st January 1923 and he appointed H. A. Ryder as its General Manager.

After acquiring the company, W. R. Morris immediately injected more capital into the business and initiated a rapid expansion. Additional premises were soon rented in George Street, Oxford, which were used for polishing, final assembly work, leak testing and offices. By 1924, production of Morris cars had reached over 500 per week and Osberton Radiators had again begun to outgrow its premises. A redundant brickworks was leased in Woodstock Road, Oxford, and a purpose-built factory erected on the site,* into which new presses and other equipment were installed. During 1926, part of this factory was occupied by The Morris Garages to produce M.G. 14/28s prior to their move to Edmund Road, and by 1939 the factory buildings had spread all over the site. (This factory is now – in 1999 – part of the Unipart Group).

When production started at the Woodstock Road factory in March 1925 there were about 500 employees and they achieved a radiator output of 1,500 per week, during that year. These radiators were for 'Bullnose' Morris Oxfords and Cowleys, 8 cwt 'Snubnose' Morris vans and Morris-Commercial 'Tonners'. The company also made the 'Joey' tinplate toy aeroplane, apparently a remnant of Doherty's business, but which served as an excellent exercise to train the workforce in soldering techniques.

While the first 'Flatnose' Morris was being made ready for the 1926 London Motor Exhibition, W. R. Morris expressed a dislike for the shape of the radiator, exclaiming that it looked like a gravestone. He requested that it be made 2" narrower and the toolroom staff at the Woodstock Road factory then set about modifying the press tools; a job that took two days and nights of continuous work, stopping only for meal breaks. W. R. Morris visited the factory just as the first radiator shell was being pressed on the modified press tools and to show his appreciation he gave the men a bonus payment of a £5 note (about £200 at 1999 values).

The factory soon became a very competitive supplier of radiators to other motor manufacturers, as well as to the Morris Group of companies, (i.e. Morris Motors Ltd., Morris Commercial Cars Ltd., The Morris Garages [for M.G.s] and from 1927, Wolseley Motors Ltd.). The company's experience in sheet metal and presswork was developed so that they were soon making such items as bonnets, petrol tanks and exhaust systems.

Following the acquisition of Osberton Radiators Ltd., by Morris Motors Ltd. in 1926, the company then became known as Morris Motors, Radiators Branch (*see page 31*) and at this time, Mr. H. A. Ryder became a director of Morris Motors (1926) Ltd. Harold Ryder also became a joint Managing Director of Morris Motors Ltd., in 1936, and the Managing Director of the M.G. Car Co., Ltd. in 1941, after Cecil Kimber had left the company, in addition to his responsibilities at Morris Motors Radiators Branch. He held these positions until he retired in 1947 when he was in his sixtieth year.

---

* The site also had access onto Bainton Road.

A 1925 season Morris 8 cwt 'Snubnose' Standard Van. The two boxes on the running board contain the battery (front) and tools

Morris 8/10 cwt Van Radiators 1924 to 1934

'Small Snubnose' radiator – pt no. 16129 – having a german silver shell, fitted to both Standard and De-luxe vans during the 1924/25/26 seasons. The dimensions around the bonnet rest flanges of pt no. 16129 and the 'Bullnose' Morris Cowley (car) radiator, pt no. 16127, are similar but 16129 has a greater cooling area. 16129 calls for bonnet boards EB2241 and EB2242 which are unique to vans because of the shape of 'Snubnose' radiator shells. Some 'Bullnose' Morris Cowley cars that were exported to countries with a hot climate were fitted with a 'small Snubnose' radiator to improve engine cooling.

'Large Snubnose' radiator fitted to 1927/28/29/30 season 8/10 cwt Light Vans. The dimensions around the bonnet rest flanges of the 'large Snubnose' radiator and the 'Bullnose' Morris Oxford (car) radiator (pt no. 16128) are similar but the 'Snubnose' radiator has a greater cooling area. Three types of this radiator were fitted as follows: – (a) Pt no. 2027, having a german silver shell, commenced at car no. 156501 (July 1926) and finished at car no. 313896 (Sept 1929), as shown. (b) Pt no. 19105 commenced at car no. 313897 (Sept 1929) and finished at car no. 322258 (Nov 1929), as shown but with a chromium-plated shell. (c) Pt no. 4949 commenced at car no. 322259 (Nov 1929) and finished at car no. 341406 (July 1930), with a chromium-plated shell and rubber mountings, which called for the deletion of the side 'feet' (as shown) and the addition of 2 studs on the base.

'Flatnose' radiator – pt no. 5486 – with a chromium-plated shell and studs for rubber mountings, fitted to 1931/32/33 season 8 cwt vans. Although a re-designed and taller radiator (pt no. 5087) was introduced for 1931 season Morris Cowley cars, with a 'winged' badge, 1931/32/33 'Flatnose' 8 cwt vans continued to be fitted with earlier style of 'Flatnose' radiator, together with a round badge. However, 'Flatnose' van radiator pt no. 5486 is not common to the chromium-plated 'Flatnose' car radiators (pt nos. 19091 and 4948) fitted during the 1930 season.

1934 season 8/10 cwt van radiator, with a chromium-plated shell pt no. 51090. This radiator is similar to that of the 1932 season Morris Cowley car but the badge on the car radiator is shield-shaped and that of the van is round.

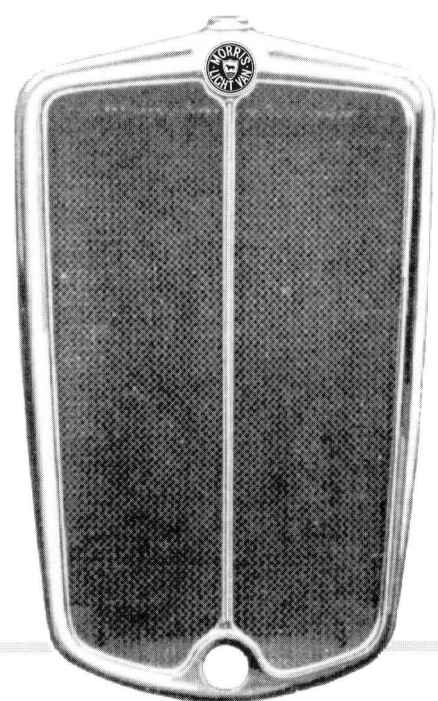

Morris 8/10 cwt Van Radiator Badges 1924 to 1934

From mid 1926, greater attention was given to names and badges in order that the products of the public company Morris Motors (1926) Ltd. (*see page 31*) would not be confused with those of the companies owned privately by W. R. Morris, such as Morris Commercial Cars Ltd. and The Morris Garages.

(1) Badge pt no. 16133

(2) Badge pt no. 17286 (below) replaced, badge pt no. 16133 (left) in late 1926 when Morris Commercial Cars Ltd. rebadged their products 'Morris-Commercial'.

(3) Badge pt no. 17286 (below) [the same pt no. as badge (2) (left)] was introduced sometime during 1930 and replaced badge (2). Badges (2) and (3) are similar except that badge (2) shows the ox in passive mood whereas in badge (3) the ox is aggressive with one front leg and his tail raised.

Radiators fitted with the above badge:
Small 'Snubnose' radiators, pt no. 16129 – german silver shell, 1924/25/26 seasons (see page 46).

Radiators fitted with the above badge:
Large 'Snubnose', pt no. 2027 – german silver shell – 1927/28/29 seasons
Large 'Snubnose', pt no. 19105 – chromium-plated shell – early 1930 season
Large 'Snubnose', pt no. 4949 – chromium-plated shell and studs for rubber mountings – 1930 season (see page 46).

Radiators fitted with the above badge:
'Flatnose', pt no. 5486 – chromium-plated shell – 1931/32/33 seasons.
8 cwt van, pt no. 51090 – chromium-plated shell – 1934 season (see page 46).

Note: Badges were soldered to radiators with german silver shells but bolted or clipped to radiators with chromium-plated shells.

Morris 8/10 cwt Van Radiator Temperature Gauges – 1924 to 1934

Above: (1) Boyce MotoMeter fitted during the 1924 and 1925 seasons.

Right: (2a & 2b) Calormeters fitted during the 1926/27/28/29/30 seasons. (a) Until late 1926 the script on the gauges read 'calometer' (i.e. no 'r'). (b) From late 1926 until July 1928, the name of the manufacturer quoted on the calormeter was 'Wilmot Birmingham', and thereafter 'Wilmot-Breeden Limited'.

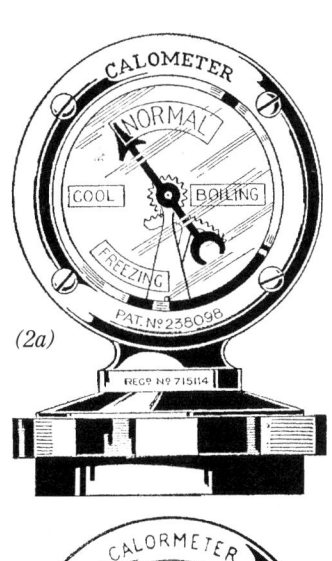

(3) Calormeter fitted during the 1931/32/33 seasons

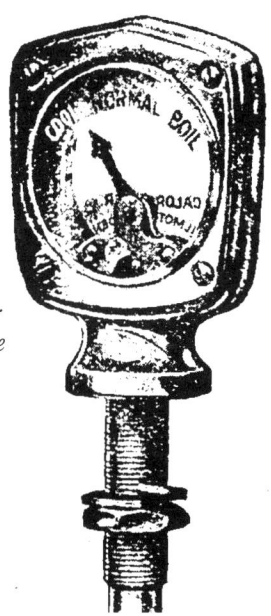

(4) Calormeter fitted during the 1934 season.

❖ 49

7

Hollick & Pratt Ltd., Coventry

During the First World War, W. R. Morris obtained the majority of bodies for the Continental-engined Morris Cowley from the well known firm of Hollick & Pratt Ltd. This company had its origins in 1876 when Edward H. Hollick bought a coachbuilding business located between Mile Lane and Quinton Road in Coventry. The business had been established in 1812 to make coaches and carriages for the nobility but, by the turn of the century, it had already begun to make bodies for motor vehicles. In about 1911, Edward Hollick decided to merge his business with another coach-building company which was jointly owned by his son-in-law, Lancelot W. Pratt, and William Fulford.* As a result, the firm of Hollick & Pratt was formed and this company was incorporated in 1922.

Athough it was W. R. Morris's policy after the First World War to assemble vehicles from bought in components, Morris Motors Ltd. erected a foundry and a bodyshop during 1919, to supplement the output of suppliers. Morris saw the need for a bodyshop because he forecast that Hollick & Pratt Ltd., would be unable to meet his requirements and that Charles Raworth & Son, Ltd., of Oxford, who had supplied most of the bodies for the 'White & Poppe' Morris

---

* W. H. Fulford probably became a director of Hollick & Pratt Ltd. in 1922 when this company was incorporated and Managing Director when Lancelot Pratt died in 1924. When Hollick & Pratt Ltd. became Morris Motors Bodies Branch in June 1926, Fulford became the General Manager of the company and also a director of Morris Motors Ltd. He held these positions until September 1927.

Right: A consignment of bodies arriving at the Cowley factory from Hollick & Pratt Ltd., Coventry, in 1925

Left: A view of the bodyshop at Cowley c.1925. M. F. (Dick) Clinch, nearest the camera, is seen with his workmate, Frank Quarterman, making a $^3/_4$ coupé body for a 'Bullnose' Morris Oxford, complete with panels and door fittings. When the men were ready to commence on another body, the foreman would issue a job card and a labourer would then bring them a set of timbers that had been rough cut in the saw mill

Oxford, were too small to make up the balance. (Following the retirement of its principal, Charles Raworth & Son, Ltd. was acquired by The Morris Garages on 2nd October 1944).

As already outlined, Morris Motors Ltd. raised £49,000 to fund their post-war expansion plans, of which £10,000 (about £400,000 at 1999 values) was in the form of a personal loan from Lancelot Pratt. By partly financing the construction of a bodyshop at Cowley to compete with his own, it might appear that Pratt was creating a conflict of interests and, to complicate matters even further, he took on the management of the new bodyshop in addition to that of his own company in Coventry.

However, Pratt controlled his own company and the Morris-owned bodyshop at Cowley as one operation with two plants. For example, four seater bodies for the Morris Cowley were made at Coventry, whereas two seaters for the same model were made at Cowley. In practice, this arrangement worked well because Morris and Pratt were close friends on both a business and a personal level. Indeed, Pratt was probably one of the best business friends that Morris ever had and the trust which Morris put in him was exceptional. When he expanded his business during the early 1920s, Morris relied heavily on Pratt for advice and support and Pratt was inevitably involved in Morris's decisions to enter into volume production of car-derived vans and purpose built commercial vehicles.

Lancelot Pratt, who was later to be recognised as a pioneer in the mass production of the wood framed/metal panelled vehicle body, believed that large scale manufacture with a comparatively small profit margin per unit was the policy for success, and Morris found that Hollick & Pratt was the only outside supplier who could produce satisfactory bodies in large quantities at competitive prices. Unfortunately, on 1st August 1922, the Hollick & Pratt factory suffered a disasterous fire. With so much wood, saw dust and wood chippings inside the premises, the fire spread quickly and soon gutted the buildings.

Before the fire, virtually all of Hollick & Pratt's output was being taken by Morris Motors Ltd. and rather than invest the insurance money back into the business, Pratt, who was by then the principal owner and Managing Director of the company, asked Morris to buy him out. A deal was struck on the spot with Morris paying £40,028 in cash for the Ordinary shares and £55,822 in instalments for the Preference shares of the company (the total of £95,850 is about £3.8 million at 1999 values). The factory was rebuilt and W. R. Morris gained control of Hollick & Pratt Ltd. officially on 1st January 1923.

Pratt then took an even more active part in Morris's business and he not only continued to run the body side but also became Morris's second-in-command, with the title of Deputy Governing Director. Tragically, these arrangements were short lived as Lancelot Pratt died of cancer on 19th April 1924 at the early age of 44, an event which deeply affected W. R. Morris (*see Profile on page 114*).

In June 1926, when Morris Motors Ltd. became a public company and absorbed W. R. Morris's personally owned companies (*see page 31*), Hollick & Pratt Ltd. became Morris Motors Bodies Branch.

Although the bodies for Morris 8 cwt vans, made between 1924 and 1926 and based on the 'Bullnose' Morris Cowley, were supplied by Davidsons of Trafford Park, Manchester, Morris 8/10 cwt vans, built from 1927 and based on the 'Flatnose' Morris Cowley, were supplied by Morris Motors Bodies Branch of Coventry. After manufacture, these bodies were transported to Morris Motors Ltd. at Cowley where they were mounted onto chassis, upholstered and finished off.

This view of the Body Mounting Shop at the Cowley factory, which was taken during 1925, includes four 'Snubnose' vans in the process of completion

8

The S.U. Company, Ltd., 1910 to 1936

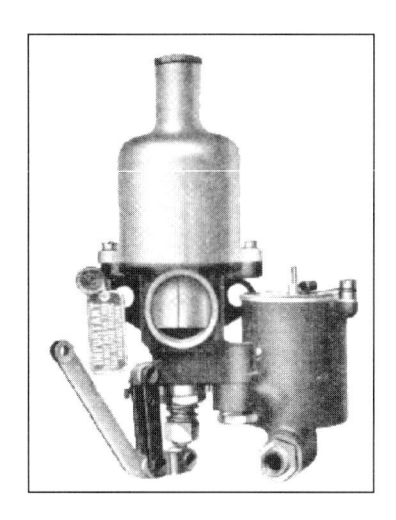

By 1900, George Herbert Skinner had three provisional patents covering his ideas about carburation. At that time, Herbert Skinner was employed by his father's shoe retailing business, Lilley & Skinner, a company that exists to this day.

Some time later Thomas Carlyle Skinner (known as Carl), who was Herbert's younger brother and who had a practical ability, had begun to try some of his brother's concepts on a Star motor car that he then owned. The idea was to place the fuel jet in an air channel that could be varied in size, in accordance with the demand of the engine, thereby giving a constant depression and air velocity. At this stage a tapered metering needle, to vary the flow of petrol, had not been thought of. Herbert was granted a full patent (no. 3257) for this device in January 1906.

T. C. Skinner

The two brothers continued to work together and their first carburetters were made at the premises of George Wailes & Co. at Euston Road, London, where Carl became a partner with George Wailes's son. In 1908 Herbert was granted another patent (no. 26,178) for a carburetter having a 'collapsible chamber' with a 'fuel needle valve' attached to it and located in an 'adjustable block' (i.e. a jet). Herbert's inventive genius had therefore devised the basic principles of the 'constant vacuum' S.U. carburetter.

Left: An S.U. type '2M' carburetter as fitted to Morris 8/10 cwt Light Vans from October 1927 until March 1930

Right: This early S.U. advertisement appeared in October 1911 and illustrates an S.U. 'Sloper' carburetter with a suction chamber enclosed by leather bellows

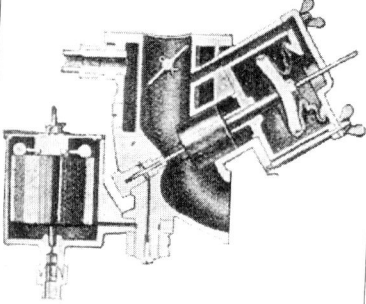

In August 1910, the S.U. Company, Ltd. (S.U. being a contraction of 'Skinners Union') was formed and some time later the company moved into premises at 154 Prince of Wales Road, Kentish Town, London. By 1913 the company's accounts showed that they were supplying 'Sloper' carburetters (so called because the suction chamber and needle assembly was positioned at an angle from the vertical, in order to reduce the fluctuations of the chamber when driving over the rough, unmetalled roads of the period) to Wolseley Motors Ltd. and the Rover Company.

During the First World War, the S.U. Company, Ltd. became engaged on munitions contracts, which included making carburetters for aero engines, with a staff of about 250. Normal production resumed after the War but the company had few customers and it showed a loss for 1919 and 1920.

Although the S.U. 'Sloper' had the advantage of automatically adjusting the flow of petrol to the engine, it was expensive and it had no proper means of providing a rich mixture for cold starting, other than 'flooding' the float chamber. This deficiency was tackled by Wolseley Motors Ltd. who added another jet and a physically operated needle. Wolseley patented their modification (no. 119187) in 1918 for which they received royalties from S.U.

When the post war boom collapsed in 1921, S.U. resorted to general engineering work, but their financial losses continued. At about this time, the S.U. 'G2' 'Sloper' was introduced which had an alloy suction chamber and piston, in place of a chamber enclosed by leather bellows, as used on the earlier type. It also had a jet which could be lowered to provide a rich mixture for cold starting. 'G2' 'Slopers' were fitted to Morris Oxfords and Cowleys during 1921/22 but, because the carburetter was mounted on the right hand side of the engine, the float chamber was located behind the jet. Consequently, when driving uphill, the fuel mixture became weaker, resulting in loss of engine power, so the 'G2' was only fitted to Morris cars for a short time.

In 1925 S.U.s introduced another type of carburetter, the '2M'. This carburetter had a similar jet arrangement to the type G2 but as its suction chamber and piston assembly were positioned vertically, the float chamber could be arranged to be on either side of the choke tube, thereby overcoming the problem with the 'G2'. The design of the '2M' is significant because it set the general pattern for S.U. carburetters thereafter. However, despite the introduction of the '2M', the S.U. Company, Ltd. continued to show financial losses, as it had done since 1919.

Up until then, the S.U Company, Ltd. had been kept afloat with loans from its five directors, all of whom were part of the Skinner family and who, in all probability, received their funds from Lilley & Skinner. Without the financial backing of Lilley & Skinner, therefore, S.U.s would almost certainly have gone out of business during the 1920s in common with many other carburetter firms of the period.

Sometime in 1926 G. H. Skinner withdrew from the S.U. company, leaving Carl to 'go it alone'. G. H. Skinner had provided nearly £8,000 (about £320,000 at 1999 values) of the £14,535 that the directors had loaned the company and he apparently had decided to give up the unequal contest, but, without his brother's help, Carl was placed in a difficult position. Towards the end of 1926 Carl, who was already known to W. R. Morris, made an unannounced visit to Morris's office at Cowley and asked him "would you like to buy the S.U. business?". When Morris asked why, Carl replied, "because it's losing money".

Throughout most of his life, W. R. Morris had been fascinated with carburetters and he often used to experiment with them, both at his factory and in his bedroom at Nuffield Place,* where a workbench and some of his tools were concealed in a wardrobe. Some time before Carl's unexpected visit, W. R. Morris had fitted an S.U. type '2M' carburetter to his own Morris

An S.U. type 'G2' Sloper carburetter

---

* W. R. Morris lived at Nuffield Place, Huntercombe, Oxfordshire from 1933, until he died in 1963. The house and its contents were inherited by Nuffield College, Oxford, who now preserve the property.

This advertisement appeared in November 1927 a few weeks after W. H. M. Burgess had been appointed as S.U.'s 'Wholesale Agent'. Note that no mention is made of the fact that the S.U. Carburetter Company was owned by W. R. Morris

S.U.

The fact that Morris Motors (1926) Ltd. are now fitting the S.U. carburetter on all their models is the finest proof one can have that it is the best carburetter in every respect. Your Morris-Oxford or Cowley, whether 1925, 1926 or 1927, can be fitted with an S.U., using all the existing controls, and the improvement in the running of your engine will amaze you. Don't delay! Have an S.U. carburetter on 30 days' free trial.

2M Type for Morris-Oxford or Morris-Cowley : **£4:10:0**

— Wholesale Agent —
W. H. M. BURGESS
36, 38, 40 Glasshouse Street
Piccadilly Circus, London, W.1
Phones : : : : : : Gerrard 8050–1686

Oxford and although he had unstinting praise for it, he did not adopt the S.U. type '2M' as original equipment on new Morris vehicles because of its prohibitive price. Both Carl Skinner and W. R. Morris had much common ground on which they could base their negotiations, and Morris later agreed to buy the S.U. Company, Ltd. for £100,000. It was also agreed that Carl should remain as manager of the business.

The first hint that Morris was getting involved with the S.U. Company, Ltd. came in November 1926 when *The Morris Owner* magazine carried an advertisement for S.U. carburetters, stating that they were 'ideal for Morris Cars' and offering them with a month's free trial.

W. R. Morris acquired the S.U. Company, Ltd. personally on the 1st December 1926 and his £100,000 payment was probably accounted for as follows:-

a. Directors loans paid off	£ 14,535 16 7
b. Settlement of the company's debts	£ 21,662 6 4
c. Purchase of the shares	£ 63,801 17 1
	£100,000 00 0[1]

1 About £4 million at 1999 values

Although the directors must have been pleased to rid themselves of their liabilities and 'get their money back', Carl apparently never wished to discuss the takeover. Perhaps he found it difficult to accept the humiliation of losing his business and then finding himself 'employed' rather than being an 'employer'.

Soon after the take over, S.U. moved from London to Birmingham where they occupied a small part of a factory at Adderley Park which W. R. Morris had just bought from the liquidators of Wolseley Motors Ltd.

The main part of this factory was soon to become the home of Morris Commercial Cars Ltd. (*see Chapter 10*) With £17,000 from W. R. Morris to purchase new plant and equipment, S.U. then prepared their new facilities for the mass production of carburetters in order to meet the demand of at least 1000 units a week from Morris Motors Ltd. alone.*

For tax reasons, W. R. Morris sold the S.U. Company, Ltd. on 27th June 1927 to his holding company, Morris Industries Ltd. who operated the business under the name of the S.U. Carburetter Company with Carl Skinner as manager. Unlike Morris's other personally-owned companies which were sold to Morris Motors Ltd. in 1926 (*see page 31*), S.U. did not become a branch of Morris Motors Ltd. as it remained, in effect, in Morris's personal ownership. This was a clever move because other motor manufacturers, who were inevitably competitors of Morris and who wished to fit S.U.s to their products, may have been unwilling to buy carburetters from a company calling itself 'Morris Motors Carburetters Branch'.

W. R. Morris did not publicise the fact that he owned S.U. and to support this he appointed W. H. M. Burgess of Glasshouse Street, London as 'Wholesale Agent' during August 1927. S.U. advertisements made no mention of the fact that W. R. Morris owned the company and, after his appointment, only quoted Burgess's name and address. (W. H. M. Burgess had been an agent for W. R. M. Motors Ltd. [*see page 26*]).

During September 1936 the S.U. Carburetter Company Ltd. was formed to take over the S.U. business from Morris Industries Ltd. and then, during the following month, Morris Motors Ltd. purchased the company (*see page 32*). Carl Skinner was then appointed as Managing Director of the new company and in addition, he became a director of Morris Motors Ltd. during November 1936. He held both of these positions until he retired in December 1947 at the age of 65.

When the Morris 'Snubnose' vans were introduced at the end of 1923 they were fitted with the Smiths single jet, type 26 HKMC, carburetters but from 1925 these vehicles were fitted with Smiths 5 jet carburetters – initially the type '4MO' and then the 'Straight Through' with a cast iron body. S.U. type '2M' carburetters were introduced on Morris 8/10 cwt Light Vans, based on the 'Flatnose' Morris Cowley, during October 1927 at car no. 223770. The S.U. type 'HV2' replaced the type '2M' in March 1930 at car no. 329750. The two types were similar; the main difference being that the type '2M' body was bronze, whereas the body of the 'HV2' was an alloy.†

Although W. R. Morris should receive great credit for realising the potential of the S.U. carburetter and for rescuing the S.U. Company, Ltd., his subsequent development of the S.U. business turned out to be one of his most underated triumphs. Morris took a small loss-making company with a technically important product and, within a short time, made it into a world-famous supplier of carburetters for both vehicles and aircraft.

---

* By 1939, over 4,000 carburetters were being produced each week.
† Morris-Commercials were fitted initially with Smiths carburetters and later Solex.

9

Specifications & Diagrams, Morris 8 & 10 cwt Light Vans by season 1924 to 1934

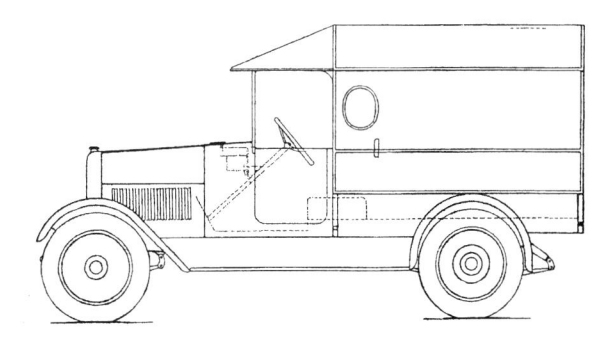

The tables that follow (*pages 64 to 92*) give general specifications for the Morris 8 and 10 cwt Light Van model types, that were manufactured between 1924 and 1934. Certain details are also shown to highlight some of the commonalities and differences between the vans and the cars from which they derived.

Although the prices of Morris 8 and 10 cwt Light Vans were reduced frequently between 1924 and 1934, as shown in the tables, their specification improved and the size of their bodies increased nearly every year.

Morris Motors Ltd. were able to achieve the steady reduction in prices primarily for two reasons: (a) the economies created by a steadily increasing level of van production and (b) because many of the components required to manufacture van chassis were derived from Morris cars. Some of these components were from cars that had already ceased to be produced when they were adopted for vans, so their design, development and tooling costs had already been accounted for. In addition, the basic design of engines, gearboxes and rear axles fitted to Morris 8 and 10 cwt Light Vans from 1924 to 1934 remained unchanged throughout the period.

Car and frame numbering and identification plates

The *car* number and the *chassis* number are the *same*. The car number is stamped on the chassis identification plate. In Morris Motors literature and in registration books the car number is often called the chassis number.

The number stamped on the r.h. chassis dumb iron is the *frame* number. Car-derived vans fitted with Hotchkiss/Morris engines have a car number that is usually 3000 higher than the frame number.

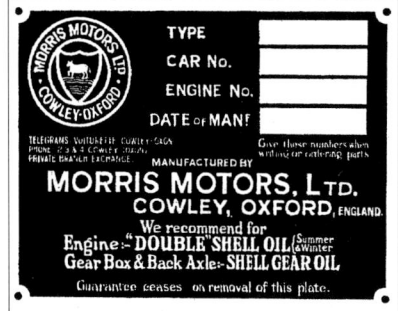

A 'Bullnose' or 'Snubnose' chassis identification plate – the 'date of manufacture' shown on the above plate was omitted during the early part of the 1925 season

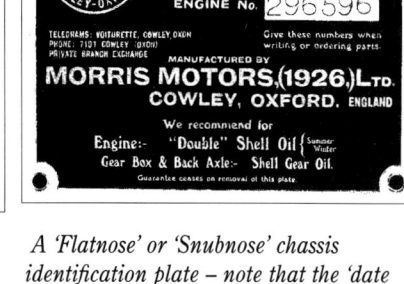

A 'Flatnose' or 'Snubnose' chassis identification plate – note that the 'date of manufacture' is not shown. (1926) was deleted from the company name in August 1929

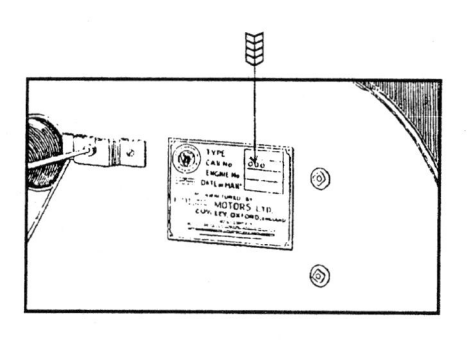

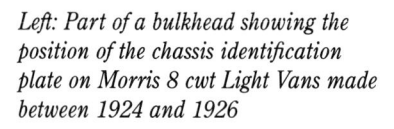

Left: Part of a bulkhead showing the position of the chassis identification plate on Morris 8 cwt Light Vans made between 1924 and 1926

Right: The bulkhead structure of Morris 8/10 cwt Light Vans made between 1927 and 1933 showing the position of the chassis identification plate

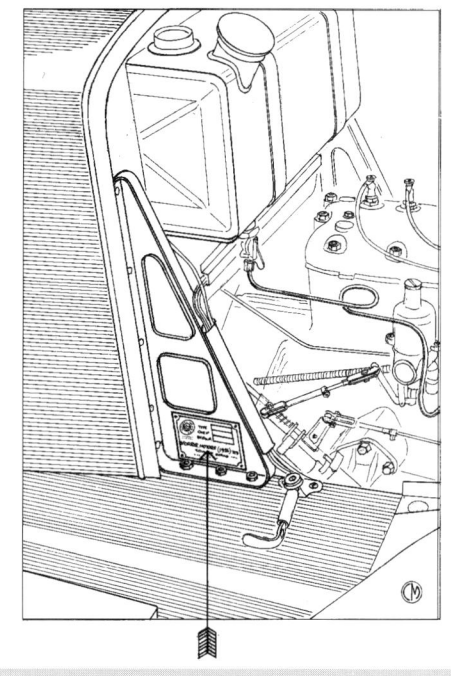

Morris 'Snubnose' 8 cwt vans, made between 1924 and 1926, were derived from the 'Bullnose' Morris Cowley and an example of a 1925 season 4 seater tourer model is shown here. The rear view of a Morris 8 cwt 'Snubnose' Standard Van can be seen in the background. Shell is alleged to have given Morris Motors Ltd. a full two gallon can of petrol (seen mounted on the running board) in consideration of displaying the words 'Use Double Shell Oil' on the engine's oil filler cap

Morris 'Snubnose' and 'Flatnose' 8/10 cwt vans made between 1927 and 1933 were derived from the 'Flatnose' Morris Cowley and an example of a 1930 saloon model, with a Kopalapso Folding Head, is shown above

Notes on the specifications and diagrams that follow

a Morris 'Snubnose' 8/10 cwt Light Vans and 'Bullnose' and 'Flatnose' Morris Oxford and Cowley cars erected between 1924 and 1934 share the same sequence of car nos. The date shown after a car no., in the following specifications, is the date that the chassis was erected.

b The script on the radiator badges, fitted to both Standard and De-luxe 'Snubnose' vans made during the 1924/5/6 seasons, reads 'Morris Commercial'. From the 1927 season, the script on radiator badges fitted to 'Snubnose' and 'Flatnose' vans reads 'Morris Light Van'. (*See page 48*).

c The mechanical components and chassis frames for 1924/5/6 'Snubnose' vans are derived from 'Bullnose' Morris Cowley cars (*see page 61*) and those for 'Snubnose' and 'Flatnose' vans made between 1927 and 1933, from 'Flatnose' Morris Cowley cars (*opposite*). When the 1933 season 'Flatnose' 8cwt van was replaced by a van derived from the then current Morris Cowley car for the 1934 season, the 'Flatnose' era ended, although 'Flatnose' van chassis continued to be made for the GPO until mid 1934.

d Because of the rounded shape of the 'Snubnose' radiator and the 'square' corners of the 'Flatnose' scuttle, the bonnets fitted to 'Snubnose' vans made between 1927 and 1930 have 'uncomfortable' curves (*see page 73*).

e The rear doors of 'Snubnose' and 'Flatnose' vans made between 1924 and 1931, have round windows with two louvres above them except for the 1924/5/6 season De-luxe Van, which has 'horizontal' oval windows with one louvre above them. The 1932 season 'Flatnose' van has 'vertical' oval windows in the rear doors with no louvres above them. The 1933 and 1934 season 8/10 cwt vans have 'vertical' oval windows in the rear doors with two louvres above them.

f 'Snubnose' and 'Flatnose' van bodies are constructed of a wood frame with 'plymax' and steel panels.

g From 1927, a steel bracing was fitted externally to the rear of 'Snubnose' and 'Flatnose' van bodies, to reinforce the base and sides of their wooden framework and joints.

h Although a headlamp mounting bar, fitted between the front wings, was introduced on Morris Cowley cars from car no. 268431 (Aug. 1928), it was not adopted on vans until the 1933 season, so 'Snubnose' and 'Flatnose' vans continued to be fitted with headlamps mounted on pillars until the end of the 1932 season.

i Some of the unladen weights quoted have been taken from data held by Morris Motors' drawing office records and may be at variance to those quoted elsewhere.

j The 1925 Morris 8 cwt 'Snubnose' van sales catalogue quotes its petrol consumption as 30 m.p.g. and its oil consumption as 1000 to 1500 m.p.g. (125 to 187.5 miles per pint). These vans were fitted with 11.9 h.p. type CB engines.

1924/25/26 season Morris 8 cwt 'Snubnose' Vans

Manufactured by Morris Motors Ltd., Cowley, Oxford.

Commenced at car no. 27412 (18th June 1923) and **finished** at car no. 156424 (28th July 1926). *Note:* 24 vans were erected at the end of the 1923 season.

Capacity. 8 cwt.

Body types/features. Type (i) Standard Van. Type (ii) De-luxe Van. Both body types were built by Davidsons, Trafford Park, Manchester, then transported to Cowley as a 'flat pack', where they were assembled, mounted and finished. The Standard Van has one side door, on the nearside, and oval side windows, whereas the De-luxe Van has two side doors and domed side windows.

Windscreen frame. 2 piece. Top frame openable. Painted black on Standard Van. Nickel plated on the De-luxe Van. Supplied by Auster or Gibson. The windscreen assemblies for the Standard and De-luxe Vans are not common.

Front wings. (i) Vans with beaded edge wheels/tyres – part no. 16003 – common to 1923 to 1926 Morris Cowley 2 & 4 strs., etc., to suit 9" wide running boards (ii) From car no: 105801 (Aug. 1925) vans with wellbase wheels/tyres – pt no. 1584 – common to 1925 Morris Oxford 2 str. and Coupé, short chassis, to suit 10" wide running boards.

Rear wings. (i) Standard Vans with beaded edge wheels/tyres – part no. 16007, bead inside – common to 1923 to 1926 Morris Cowley 2 strs., etc. (ii) De-luxe Vans with beaded edge wheels/tyres – part no. 16010, bead inside – common to 1924 to 1926 Morris Cowley 4 strs., etc. (iii) From car no. 105801 (Jul. 1925) vans with wellbase wheels/tyres – part no. 1528 – common to 1925 to 1926 Morris Oxford 4 str., etc.

Chassis frame.* Pt no. 1511, a frame of deeper section was introduced in 1925 and a bar between the front dumb irons was introduced in 1926.

Engine. Type CB – 11.9 h.p. (*see page 40*).

Carburetter.* 1924 season: Smiths single jet type 26HKMC. Early 1925 season: Smiths 4MO 5 jet. Late 1925 season and 1926 season: Smiths 'Straight-through' 5 jet.

Transmission.* 3 speed gearbox. Prop. shaft enclosed by torque tube.

Steering box.* Worm and wheel.

Brakes. Rear wheel only, standard. Rod operated.

Road springs. *Front* (i) NFWB 6 leaves, part no. 1276* (ii) FWB 6 leaves + 2 rebound, part no. 16399*. *Rear* ³/₄ elliptic with 9 leaves on lower spring, pt no. 1453, 7 leaves on upper spring, pt no. 1420.

Rear axle ratio. 4.75:1.

Radiator. 'Small' Snubnose, pt no. 16129 (*see page 46*).

Unladen weight (1924 season).
Standard van   17 cwt. 3 qts. 0 lbs.
De-luxe van   18 cwt. 0 qts. 0 lbs.

Prices.

	1924 season	1925 season	1926 season
Standard	£198.0.0	£180.0.0	£167.0.0
De-luxe	£220.0.0	£200.0.0	£192.10.0

Wheelbase. 8' 6".   **Track.** 4' 0".

---

* Common to 'Bullnose' Morris Cowleys according to season

The interior of a 1925 season Morris 8 cwt 'Snubnose' Standard Van

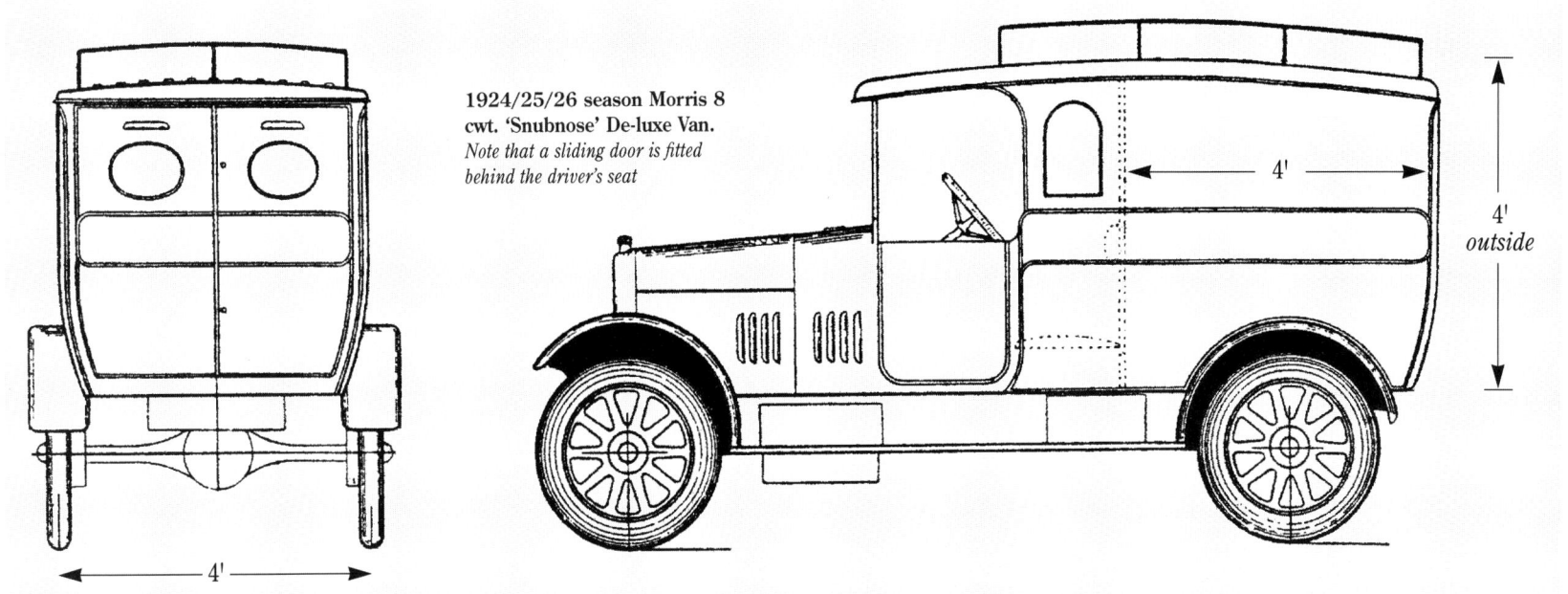

1924/25/26 season Morris 8 cwt. 'Snubnose' De-luxe Van.
Note that a sliding door is fitted behind the driver's seat

Notes
(1) 1924 season vans have a 3 lamp set with combined head and side lamps mounted on the front wings.
(2) 1925 and 1926 seasons vans have a 5 lamp set with head lamps mounted on pillars that are bolted to the chassis frame.
(3) Wellbase wheels/tyres replaced beaded edge wheels/tyres at car no. 105801 (Jul. 1925)

1924/25/26 season Morris 8 cwt 'Snubnose' Vans

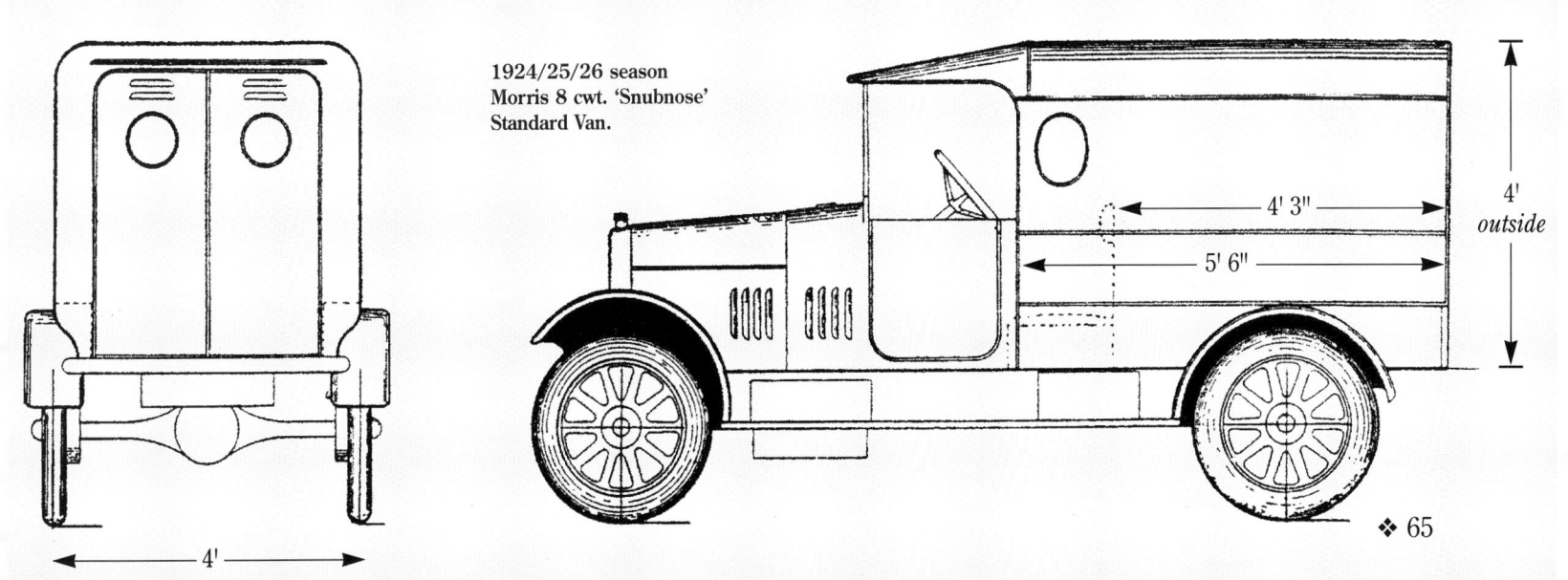

1924/25/26 season Morris 8 cwt. 'Snubnose' Standard Van.

❖ 65

1924 season 'Snubnose' 8 cwt Standard Van

Fitted with a 3 lamp set, having combined head and side lamps mounted on the front wings

The Morris Commercial Standard Van

Double doors at back of body, double adjustable windscreen, single door on near side to driver's seat, dynamo lighting set, three lamps, speedometer, oil gauge, spring gaiters, bulb horn, spare petrol can and carrier, licence holder, complete kit of tools.   Finished in two coats lead colour, two coats of filling, and rubbed down ready for painting and lettering ... ... Price – £198

To carry a net load of 8 cwt.

1924 season 'Snubnose' 8 cwt De Luxe Van

Fitted with a 3 lamp set, having combined head and side lamps mounted on the front wings

The Morris Commercial Van de Luxe

Double doors at back of body, and sliding doors behind driver's seat. Two side doors in front so that the driver can enter from either side. Double adjustable windscreen with plated metal work. Electric lighting set. Speedometer and Oil Gauge, spare Petrol Can and Carrier, Bulb Horn, Spring Gaiters and Licence Holder, complete Kit of Tools. Finished in two coats of lead colour, two coats of filling, and rubbed down ready for painting and lettering   ...   ...

Price – £220

To carry net load of 8 cwt.

1925 season
'Snubnose' 8 cwt
Standard Van

Price – £180

1925 season
'Snubnose' 8 cwt
De Luxe Van

Price – £200

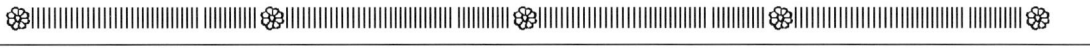

1926 season 'Snubnose' 8 cwt Standard Van
Price – £167

1926 season 'Snubnose' 8 cwt De Luxe Van
Price – £192.10.0

The driver's door is openable on De Luxe Vans

1927 season Morris 8 & 10 cwt 'Snubnose' Light Vans

Manufactured by Morris Motors Ltd., Cowley, Oxford

Commenced car no. 156501 (July 31st 1926), **Finished** Car no. 215000 (August 22nd 1927). Note: car nos. 156425 to 156500 inclusive were not used.

Capacity. Type (i) 8 cwt, 58.5 cu ft (ii) 10 cwt, 75 cu ft.

Body types/features. Type (i) 8 cwt Van. Type (ii) 10 cwt Van with a 'High Top'. Both types designed at Cowley and built at Coventry (Bodies Branch) and then mounted and finished at Cowley. Type (ii) is higher, wider and longer than type (i), but the size of the rear doors on both types is similar. Type (ii) probably replaced type (i) during the middle of the 1927 season. Both types of van are adapted from the 1924/25/26 Standard Van with modifications to suit the 'Flatnose' type dash/scuttle and chassis frame, which has an 'arch' over the rear axle (unlike the 'Bullnose' chassis frame). Oval windows in the side panels. The side doors are not fitted with either windows or side screens.

Windscreen frame. 2 piece. Top frame openable.

Front wings. Pt no. 2216 – unique to vans – the flat section at the base of wing 2216 is approx. $4\frac{1}{2}$" wide (*see photograph*), whereas this dimension on the front wings of 1927 and 1928 'Flatnose' Morris Cowleys is approx. $3\frac{1}{2}$". Wing 2216 is to suit the 'Snubnose' radiator which is about 2" narrower than a 'Flatnose' Morris Cowley radiator.

Rear wings. Pt no. 2000 – common to 1927 season Morris Cowley 4 str., etc.

Chassis frame. Pt no. 1849 – common to 1927/28 season 'Flatnose' Morris Cowley.

Engine. Type CB. 11.9 h.p. (*see page 40*).

Carburetter.* Smiths 'Straight-through' 5 jet.

Transmission.* 3 speed gearbox. Prop shaft enclosed by torque tube.

Steering box.* Worm and wheel.

Brakes. Rear wheel only standard, rod operated.

Road springs. *Front*, 6 leaves + 2 rebound, pt no. 1922*. *Rear*, 8 leaves, pt no. 2155 (unique to vans).

Rear axle ratio. 4.75:1.

Radiator. 'Large Snubnose' – pt no. 2027 (*see page 46*).

Prices. Type (i) £160.00. } Simplified
Type (ii) probably £160.00 } models

Note: Front wheel brakes and full Cowley equipment £12.10.0 extra.

Unladen Weight. Type (i) = 17 cwt. 3 qtrs. 0 lbs
Type (ii) = n/a

Wheelbase. 8' 9".

Track. 4'.

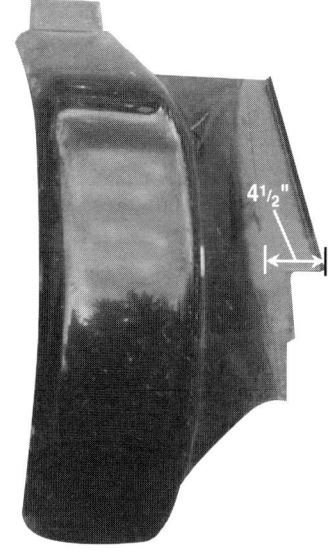

Front wing, pt no. 2216.

---

* Common to 1927 season 'Flatnose' Morris Cowley.

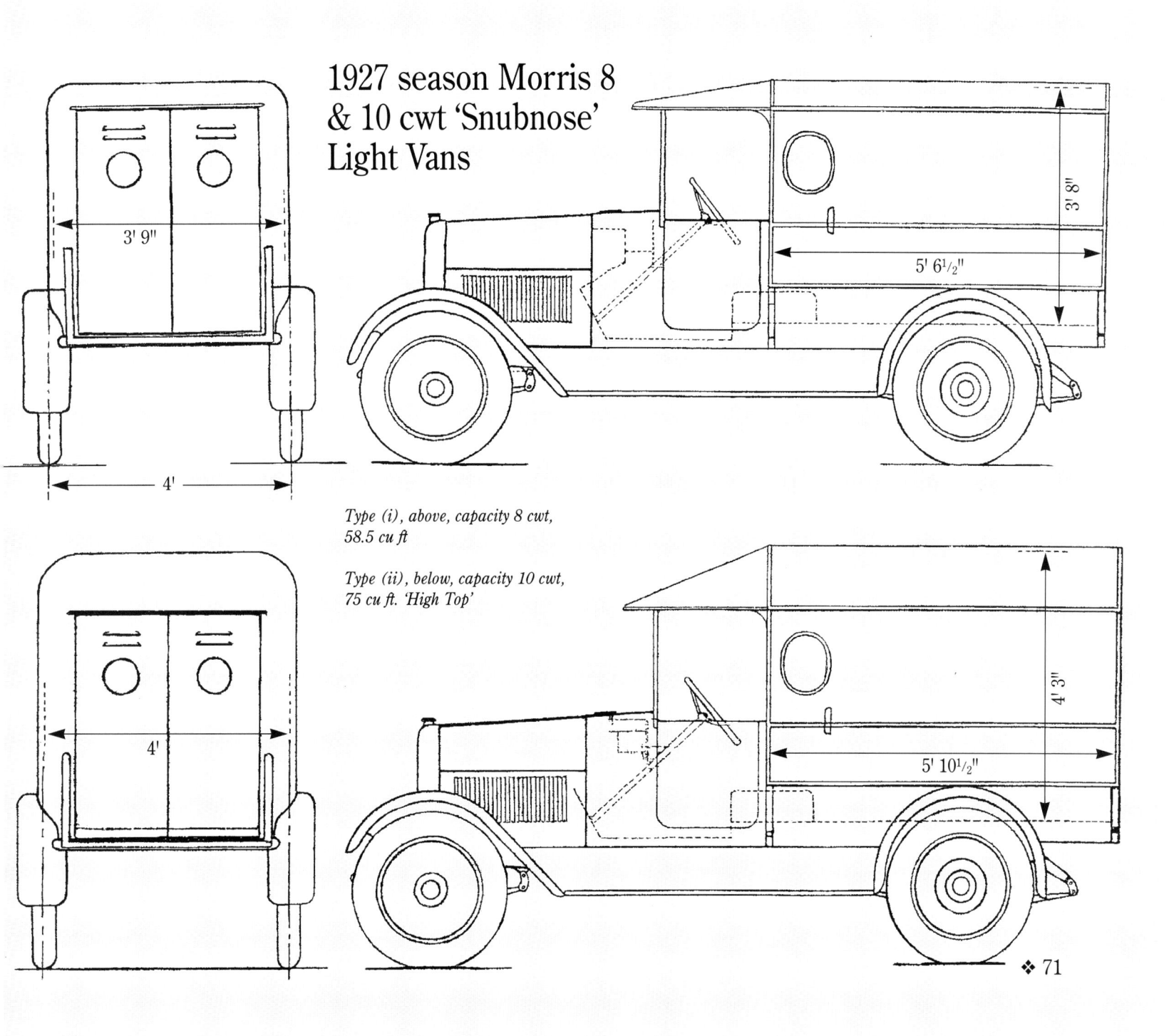

1927 season Morris 8 & 10 cwt 'Snubnose' Light Vans

Type (i), above, capacity 8 cwt, 58.5 cu ft

Type (ii), below, capacity 10 cwt, 75 cu ft. 'High Top'

❖ 71

1927 season Morris 8 & 10 cwt 'Snubnose' Light Vans

A 1927 season Morris 10 cwt 'High Top' (Type ii) Light Van, pictured in the Despatch Department at the Cowley factory

1927 season Morris 8 cwt Light Van (Type i). Note the 'uncomfortable' shape of the bonnet, resulting from the marriage of the rounded shape of the radiator to the sharper corners of the 'Flatnose' derived bulkhead

1927 season Morris 8 & 10 cwt 'Snubnose' Light Vans

❖ 73

1928 season Morris 10 cwt 'Snubnose' Light Vans

Manufactured by Morris Motors Ltd., Cowley, Oxford

Commenced at car no. 215001 (22nd August 1927).
Finished at car no. 268430 (21st August 1928).

Capacity. 10 cwt. 75 cu ft.

Body type/features. Similar to the 1927 'High Top' Light Van but with an enclosed drivers cab, as the side doors now have winding windows. Designed at Cowley and built at Coventry (Bodies Branch) and then mounted and finished at Cowley. The rear doors are higher when compared with the 1927 season Light Vans, to take full advantage of the high roof. Oval windows in the side panels.

Windscreen/frame. Single piece openable. Frame is integral with the bodywork.

Front wings. Pt no. 2216, as 1927 season.

Rear wings. Pt no. 1999 – common to 1927/8 season Morris Cowley 2 strs. and 1928 season Morris Cowley 4 strs., etc.

Chassis frame. Pt no. 1849, as 1927 season.

Engine. Type CB, 11.9 h.p. (*see page 40*).

Carburetter.* Initially Smiths 'Straight-through' 5 jet. From car no. 223770, SU type 2M.

Transmission.* 3 speed gearbox. Prop. shaft enclosed by torque tube.

Steering box.* Worm and wheel.

Brakes. Four wheel brakes became standard during the season – rod operated.

Road springs. *Front* – initially pt no. 1922, as 1927 season. From car no. 232634, pt no. 2694 with 5 leaves + 2 rebound (with FWB).*

Rear pt no. 2155 as 1927 season.

Rear axle ratio. 5:1.

Radiator. 'Snubnose' – pt no. 2027 (*see page 46*). As 1927 season.

Price. £165.0.0 with FWB and full Cowley equipment.

Unladen weight. 20 cwt. 1 qtr. 0 lbs.

Wheelbase. 8' 9".

Track. 4' 0".

---

* Common to 1928 season 'Flatnose' Morris Cowleys.

This van is similar to the 1927 season 10 cwt High Top (page 72) but the side doors now have windows and the rear doors are higher

1928 season Morris 10 cwt 'Snubnose' Light Vans

1929 & 1930 season Morris 10 cwt 'Snubnose' Light Vans

Manufactured by Morris Motors Ltd., Cowley, Oxford

Commenced at car no. 268431 (21st August 1928) and finished at car no. 341406 (18th July 1930).

Capacity. 10 cwt 78.75 cu ft.

Body type/features. Similar to the 1928 season 10 cwt Light Van except the 'steps' in the side panels are deleted so that the side panels are now 'flat' to the floor and the rear wings/wheel arches are now integral with the body. Rear doors are increased in width to take advantage of the wider lower section of the body and floor. Oval windows in side panels. The semi-circular wooden protrusions at the base of each corner (*not shown on the drawing opposite but seen on pages 78 & 79*) of the body were deleted from the 1930 season when the construction of the body was lightened.

Windscreen/Frame. Single piece openable. Frame is integral with bodywork.

Front wings. (i) Initially pt no. 2216 as 1927 season. (ii) From car no. 289817 (Jan. 1929) pt no. 4084 o/s and 4085 n/s – similar to (i) but have a cut out to suit rubber engine mountings – finished at car no. 322258 (Nov. 1929), unique to vans. (iii) From car no. 322259 (Nov. 1929), pt no. 4990 o/s and 4991 n/s – similar to (ii) but base modified to suit rubber mounted radiator – finished at car no. 345891 (Oct. 1930), unique to vans.

Rear wings. Pt no. 3422, integral with body, dia. approx. 38" (i.e. about 4" larger dia. than cars), unique to vans.

Chassis frame. (i) Initially pt no. 1849 as 1928 season vans. (ii) From car no. 289817 (Jan. 1929) pt no. 4076 to suit rubber engine mountings.

Engine. Type CB, 11.9 h.p. (*see page 40*). Rubber engine mountings introduced at car no. 289817 (Jan. 1929) with bonnet pt no. 4083 to suit.

Carburetter.* Initially, SU Type 2M. From car no. 329750 (Mar. 1930) SU Type HV2.

Transmission.* 3 speed gearbox. Gear lever spring-loaded from engine no. 371263 (Apr. 1930). Prop. shaft enclosed by torque tube.

Steering box. Worm & wheel.

Brakes. Four wheel brakes standard, rod operated.

Road springs. *Front* pt no. 2694, as late 1928 season. *Rear* pt no. 2155 as 1927/28 season.

Rear axle ratio. 5:1.

Radiators. (i) Initially pt no. 2027, as 1928 season. (ii) Chromium-plated 'Snubnose' radiator, pt no. 19105 – fitted from car no. 313897 (Sept. 1929) until car no. 322258 (Nov. 1929). (iii) Chromium-plated 'Snubnose' radiator – pt no. 4949 – rubber mounted, fitted from car no. 322259 (Nov. 1929) until car no. 341406 (July 1930). (*See page 46*).

Price. 1929 & 1930. £165.0.0 with FWB & full Cowley specification.

Unladen Weight. 1929 = 20 cwt. 0 qtrs. 4 lbs.
1930 = 19 cwt. 3 qtrs. 2 lbs.

Wheelbase. 8' 9". **Track.** 4'.

* Common to 1929/30 season 'Flatnose' Morris Cowleys.

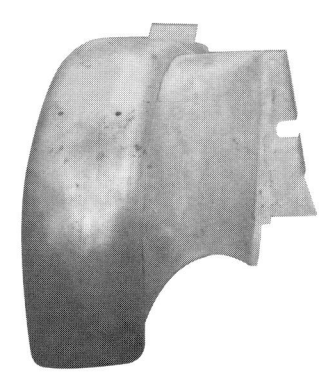

Front wings pt nos. 4990 O/S (above) & 4991 N/S (below)

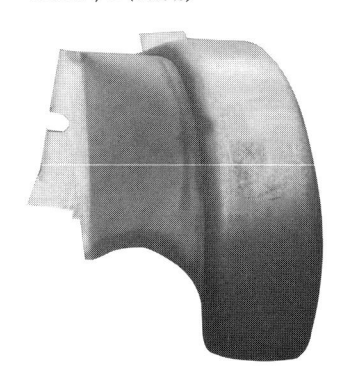

Rear wing, pt no. 3422

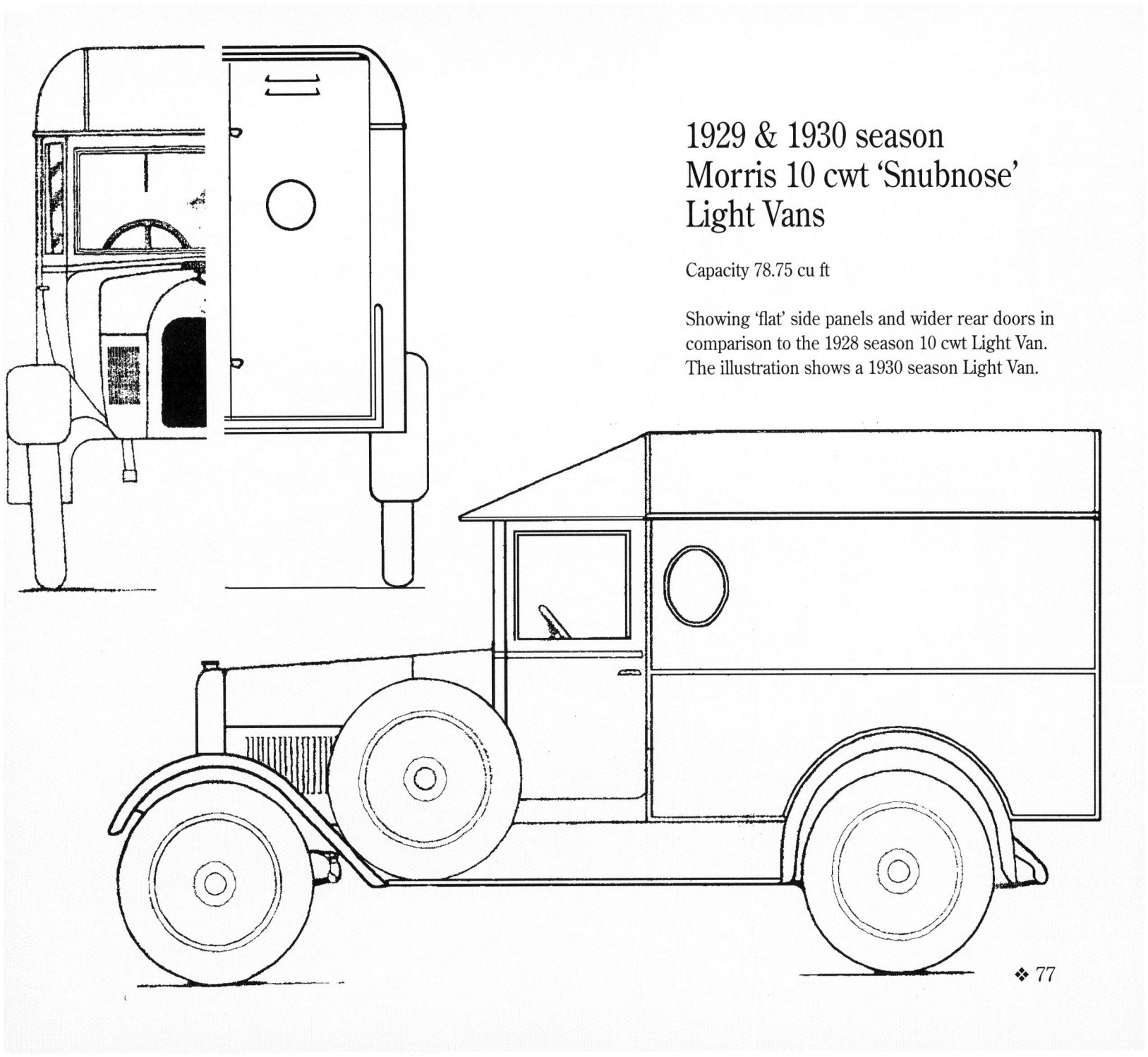

1929 & 1930 season Morris 10 cwt 'Snubnose' Light Vans

Capacity 78.75 cu ft

Showing 'flat' side panels and wider rear doors in comparison to the 1928 season 10 cwt Light Van. The illustration shows a 1930 season Light Van.

1929 & 1930 season Morris 10 cwt 'Snubnose' Light Vans

The 11.9 h.p. Morris Light Van

TO the tradesman of to-day, speedy, reliable and economical motor transport is an absolute necessity. Time has proved that in the light transport class no vehicle exists which is superior to the Morris Light Van in serviceability or profit-earning capacity. With its sturdy chassis, powerful engine, excellent springing, and generous tyre equipment, it can carry a full load at an attractive speed and with exceptional smoothness. Requiring the minimum of attention, it is capable of continuous service over long periods, and is in every way dependable transport.

The body of the 1929 Morris Light Van has been yet further improved, its internal dimensions having been increased to 4 ft. 9 in. by 4 ft. by 4 ft. 3 in., giving a cubic capacity of 78¾ cubic feet, the exceptionally large rear doors ensuring ease of loading. Access to the driving compartment is achieved by a large door on each side. Winding windows to the doors and a single-panel adjustable windscreen provide complete protection for the driver.

It is delivered with complete equipment and finished in shop grey, enabling the purchaser to choose whatever colour finish suits his individual requirements.

The equipment includes:—

Speedometer, clock, oil gauge, petrol gauge (in tank), two-level petrol tap, automatic windscreen wiper (black finish), pressure lubricating pump, licence holder, calormeter and wings, driving mirror, shock absorbers, spring gaiters, electric horn, dash-operated ventilator, electric lighting and starting, five-lamp equipment, dashlamp, nickel-plated hub nuts, tin of lubricating oil, five detachable steel wheels, five Dunlop heavy service tyres, toolbox and tool kit, spare wheel carrier.

Price £165

QUALITY · VALUE · ECONOMY · SAFETY

The illustrations left and opposite show 1929 season Light Vans, having semi-circular wooden protrusions at the base of the body which were deleted from the 1930 season

1929 & 1930 season Morris 10 cwt 'Snubnose' Light Vans

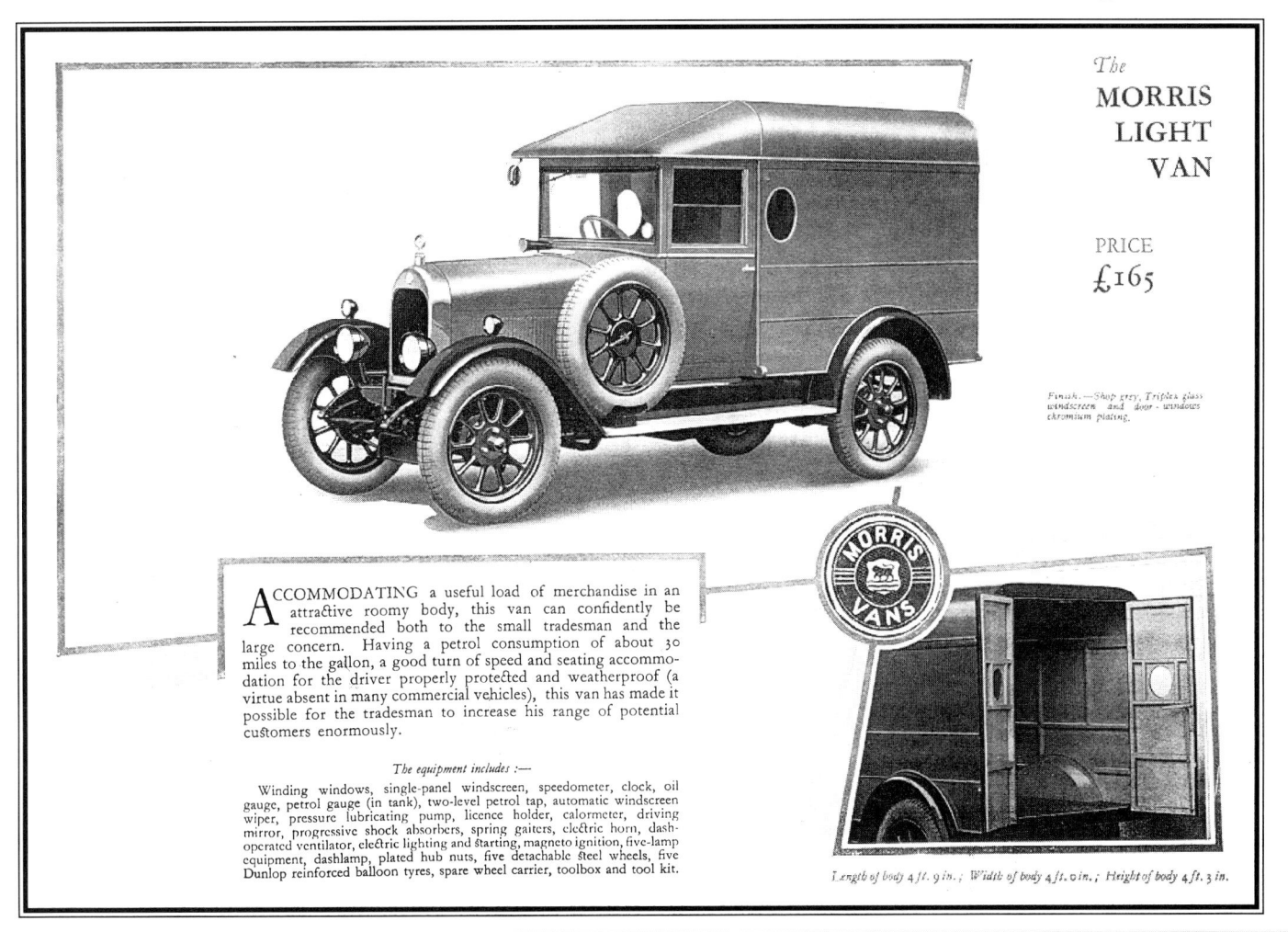

The
MORRIS
LIGHT
VAN

PRICE
£165

Finish.—Shop grey, Triplex glass windscreen and door-windows chromium plating.

ACCOMMODATING a useful load of merchandise in an attractive roomy body, this van can confidently be recommended both to the small tradesman and the large concern. Having a petrol consumption of about 30 miles to the gallon, a good turn of speed and seating accommodation for the driver properly protected and weatherproof (a virtue absent in many commercial vehicles), this van has made it possible for the tradesman to increase his range of potential customers enormously.

The equipment includes :—

Winding windows, single-panel windscreen, speedometer, clock, oil gauge, petrol gauge (in tank), two-level petrol tap, automatic windscreen wiper, pressure lubricating pump, licence holder, calormeter, driving mirror, progressive shock absorbers, spring gaiters, electric horn, dash-operated ventilator, electric lighting and starting, magneto ignition, five-lamp equipment, dashlamp, plated hub nuts, five detachable steel wheels, five Dunlop reinforced balloon tyres, spare wheel carrier, toolbox and tool kit.

Length of body 4 ft. 9 in.; Width of body 4 ft. 0 in.; Height of body 4 ft. 3 in.

1931 season Morris 8 cwt 'Flatnose' Light Vans

Manufactured by Morris Motors Ltd., Cowley, Oxford

Commenced at car no. 341407 (28th July 1930).

Finished at car no. 358192 (17th July 1931).

Capacity. 8 cwt, 78 cu ft.

Body types/features. Adapted from the 1930 season Light Van body, but a curvature is added to the roof above the driver. Designed at Cowley and built at Coventry (Bodies Branch) and then mounted and finished at Cowley.

Windscreen/frame. Single piece, openable. Frame is integral with the bodywork.

Front wings. (i) Initially pt no. 4990 n/s & 4991 o/s as 1929/30 season, type (iii). (ii) From car no. 345892 (Oct. 1930) pt no. 5111 o/s and 5112 n/s – unique to vans.

Rear wings. Pt no. 3422 as 1929/30 season.

Chassis frame. (i) Initially pt no. 4076 as late 1929 & 1930 season. (ii) From car no. 345892 (Oct. 1930) pt no. 5104 – to suit introduction new type wings, etc.

Engine. Type CG 14/32, 13.9 h.p., with 'turbulent' (Ricardo) cylinder head – introduced at eng no. 384228 (*see page 40*). The first van to have a type CG engine and chromium-plated 'Flatnose' radiator was car no. 341532 (July 1930). Engines for export vans were fitted with a water pump, from car no. 342898 (Sept. 1930).

Carburetter.* SU Type HV2.

Transmission.* 3 speed gearbox. Gear lever spring-loaded. Prop. shaft enclosed by torque tube.

Steering box. Initially worm and wheel. Bishop Cam from car no. 346061 (Oct. 1930) for vans.†

Brakes. Four wheel brakes standard – rod operated.

Road springs. *Front*, initially pt no. 2694 as late 1928, 1929/30 season. Pt no. 3307, with 5 leaves, from car no. 345892 (Oct. 1930) – common to 1929/30 season Morris Cowleys. *Rear*, pt no. 2155 as fitted from 1927 season.

Rear axle ratio. 5:1.

Radiator. Pt no. 5486, chromium-plated 'Flatnose' with rubber mountings – requires bonnet pt no. 18343. (The first van to have a chromium-plated 'Flatnose' radiator was car no. 341532 (July 1930). (*See page 47*).

Price. £160.0.0 with FWB and full Cowley equipment.

Unladen weight. 19 cwt. 3 qtrs. 26 lbs.

Wheelbase. 8' 9".

Track. 4'.

---

* Common to 1931 season 'Flatnose' Morris Cowleys.

† Bishop Cam steering introduced for 'Flatnose' Morris Cowleys at car no. 326301 (Jan 1930).

1931 season Morris 8 cwt 'Flatnose' Light Vans

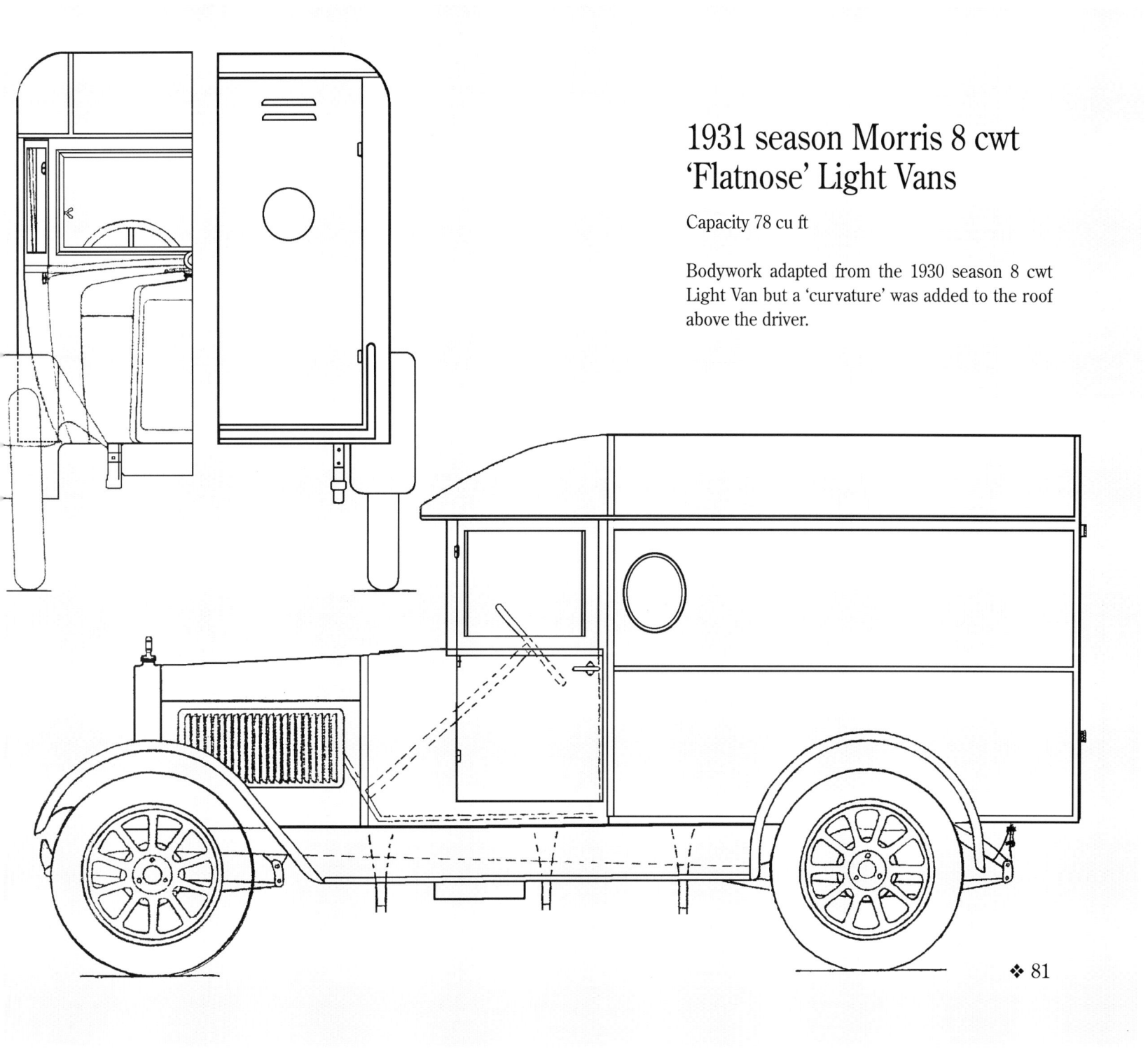

Capacity 78 cu ft

Bodywork adapted from the 1930 season 8 cwt Light Van but a 'curvature' was added to the roof above the driver.

1931 season Morris 8 cwt 'Flatnose' Light Vans

The MORRIS LIGHT VAN
(8-cwt.)

EVERY progressive tradesman fully appreciates the value of fast, reliable and economical light transport, and the importance of the extended trading scope which it offers him. The Morris Light Van has definitely proved itself to be unapproachable in its class and capable of that absolute reliability which is so essential to the trader.

For 1931 a more powerful engine of 14/32 h.p. is fitted, giving quick acceleration for fast traffic work and excellent hill-climbing with full load of 8 cwt.

Provided with a spacious body measuring 4 ft. in width, 4 ft. 9 in. in length, 4 ft. 3 in. in height, and possessing a cubic capacity of 78 cubic feet, a sturdy chassis unit, excellent springing, and an efficient engine, it can transport its full useful load of 8 cwt. at an attractive speed smoothly and rapidly. Loading ease is assured by large double doors at the rear, and access to the seats is achieved through a wide door on each side.

EQUIPMENT. Winding windows in doors, single-panel windscreen, speedometer, clock, oil gauge, petrol gauge (in tank), two-level petrol tap, automatic wind-screen wiper, pressure chassis lubricating pump, licence holder, calormeter, driving mirror, progressive shock absorbers, spring gaiters, electric horn, dash-operated ventilator, electric lighting and starting, magneto ignition, five-lamp equipment, dashlamp, five detachable steel wheels, five Dunlop reinforced cord balloon tyres, spare wheel carrier, jack, tyre pump, toolbox and full kit of tools.

Price . . . £160

COACHWORK. The Morris Light Van is delivered finished in shop grey, with Triplex safety glass windscreen and winding door-windows, chromium finish.

LIGHT VAN CHASSIS (equipment as for 11.9 h.p. Morris-Cowley) . . £125 ex Works.

A Morris 8 cwt 'Flatnose' Light Van chassis, with equipment (i.e. wings, headlamps, headlamps mounting bar and bumpers) as for a 1931 season Morris Cowley

1931 season Morris 8 cwt 'Flatnose' Light Vans

1932 season Morris 8 cwt 'Flatnose' Light Vans

Manufactured by Morris Motors Ltd., Cowley, Oxford

Commenced at car no. 358193 (July 1931). **Finished** at car no. 373499 (July 1932).

Note: 'Flatnose' Morris Cowley cars ceased to be manufactured at the end of the 1931 season, but 'Flatnose' vans continued to be made.

Capacity. 8 cwt, 78 cu ft.

Body types/features. Adapted from the 1931 season Light Van body, but the side doors are increased in width, the windows in the side panels are deleted, oval windows are put into the rear doors in place of round ones and the louvres in the rear doors are deleted. Designed at Cowley and built at Coventry (Bodies Branch) and then mounted and finished at Cowley.

Windscreen/frame. As 1931 season.

Front wings. Pt nos. 50116 (o/s) and 50117 (n/s) – unique to vans.

Chassis frame. Pt no. 5104, as 1931 season.

Rear wings. Pt nos. 50118 (o/s) and 50119 (n/s) – unique to vans.

Engine. Type CG 14/32, 13.9 h.p. (*see page 40*), fitted with a 'turbulent' type (Ricardo) cylinder head and coil ignition (Lucas DC4 unit). Water pump fitted to export models.

Carburetter. SU Type HV2.

Transmission. 3 speed gearbox. Gear lever spring-loaded. Prop shaft enclosed by torque tube.

Steering box. Bishop Cam.

Brakes. Four wheel brakes standard, rod operated.

Road springs. *Front* pt no. 3307, as late 1931 season. *Rear* pt no. 2155, as fitted from 1927 season.

Rear axle ratio. (i) 5:1 until car no. 362011 (ii) 5.27:1 from car no. 362012.

Radiator. Pt no. 5486, as 1931 season (*See page 47*).

Price. £160.

Unladen weight. 20 cwt. 1 qtr. 10 lbs.

Wheelbase. 8' 9".

Track. 4'.

1932 season Morris 8 cwt 'Flatnose' Light Vans

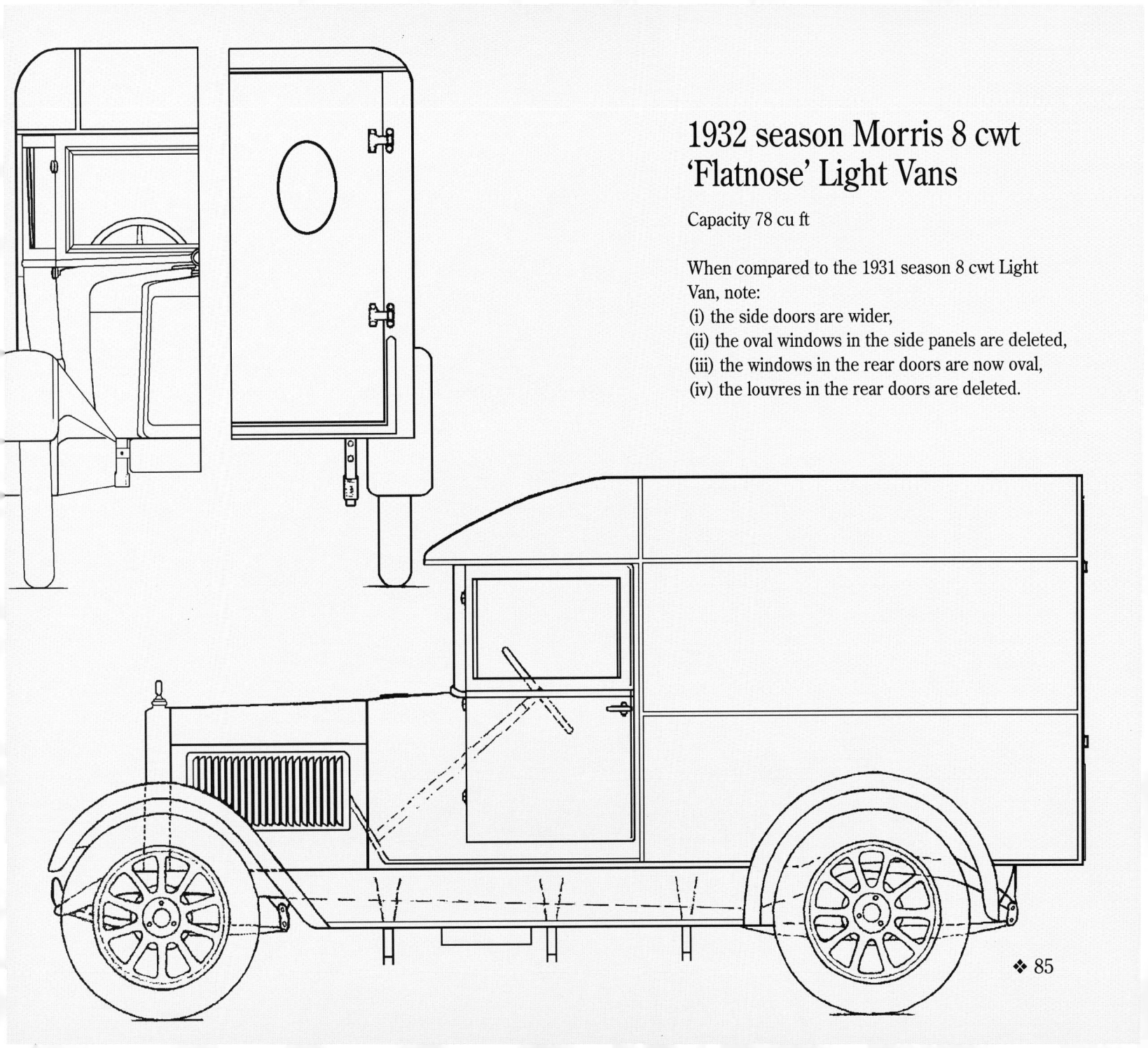

Capacity 78 cu ft

When compared to the 1931 season 8 cwt Light Van, note:
(i) the side doors are wider,
(ii) the oval windows in the side panels are deleted,
(iii) the windows in the rear doors are now oval,
(iv) the louvres in the rear doors are deleted.

1932 season Morris 8 cwt 'Flatnose' Light Vans

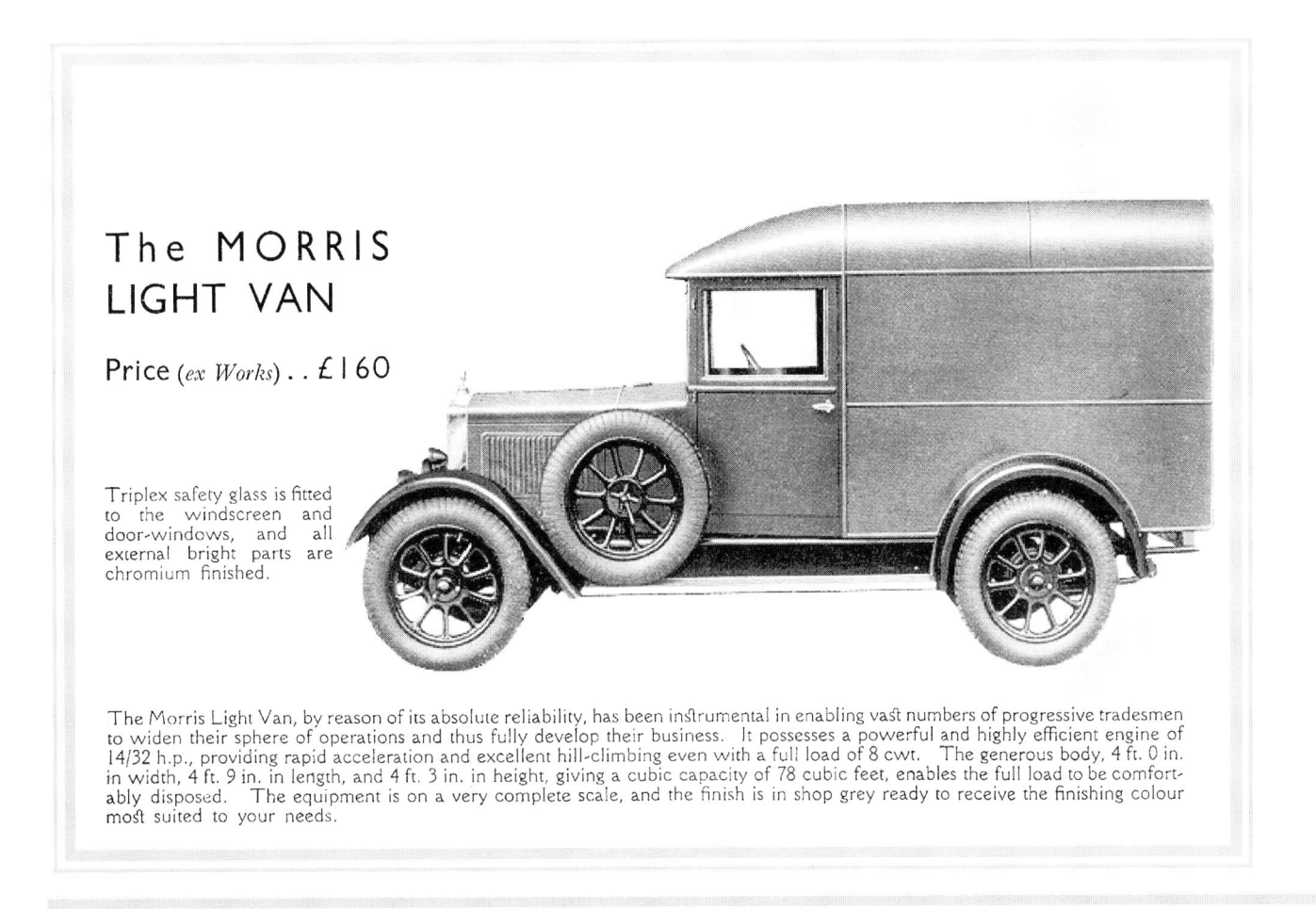

1932 season Morris 8 cwt 'Flatnose' Light Vans

The MORRIS LIGHT VAN

Price *(ex Works)* .. £160

Triplex safety glass is fitted to the windscreen and door-windows, and all external bright parts are chromium finished.

The Morris Light Van, by reason of its absolute reliability, has been instrumental in enabling vast numbers of progressive tradesmen to widen their sphere of operations and thus fully develop their business. It possesses a powerful and highly efficient engine of 14/32 h.p., providing rapid acceleration and excellent hill-climbing even with a full load of 8 cwt. The generous body, 4 ft. 0 in. in width, 4 ft. 9 in. in length, and 4 ft. 3 in. in height, giving a cubic capacity of 78 cubic feet, enables the full load to be comfortably disposed. The equipment is on a very complete scale, and the finish is in shop grey ready to receive the finishing colour most suited to your needs.

1933 season Morris 8 cwt 'Flatnose' Light Vans

Manufactured by Morris Motors Ltd., Cowley, Oxford

Commenced at car no. 373500 (July 1932). **Finished** at car no. 382397 (July 1933)*.

Capacity. 8 cwt. 70 cu ft.

Body types/features. New body design introduced. Designed at Cowley and built at Coventry (Bodies Branch) and mounted and finished at Cowley.

Windscreen/frame. Integral with body, openable and is raked.

Front wings. Pt nos. 50801 (o/s) & 50802 (n/s) unique to vans. A headlamp mounting bar is fitted between the wings.

Rear wings. Pt no. 3637 (o/s) & 3638 (n/s) – common to 1929 season Morris Cowley 2 & 4 seaters.

Chassis frame. Pt no. 3521, common to 1929/30 season Morris Cowleys.

Engine. Type CN, 13.9 h.p. (*see page 40*). Chain driven camshaft, 'turbulent' type (Ricardo) head, coil ignition with distributor driven from camshaft. Car type fume consumer not fitted. From eng no. 422384 each clutch plate has 36 corks, in lieu of 26 as previously – flywheel, etc., modified to suit.

Carburetter. SU Type HV2.

Transmission. 3 speed gearbox similar to earlier types but laygear is supported by caged needle roller bearings, in lieu of bushes. Gear lever spring-loaded. Prop. shaft enclosed by torque tube.

Steering box. Bishop Cam.

Brakes. Four wheel brakes, rod operated.

Road springs. *Front* pt no. 3307, as fitted to 1931/32 season cars and vans. *Rear* pt no. 2155, as fitted from 1927 season vans.

Rear axle ratio. 5.27:1.

Radiator. Pt no. 5486, as 1931/32 season. (*See page 47*).

Price. £160.0.0 (chassis complete with lamps, bonnet, wings and running boards – £125.00).

Unladen weight. 20 cwts. 1 qtr. 10 lbs.

Wheelbase. 8' 9".

Track. 4'.

---

* Although 'Flatnose' Morris Light Vans were withdrawn from general sale at the end of the 1933 season, 164 'Flatnose' van chassis were erected during the 1934 season specially for the GPO (*see page 141*).

1933 season Morris 8 cwt 'Flatnose' Light Vans

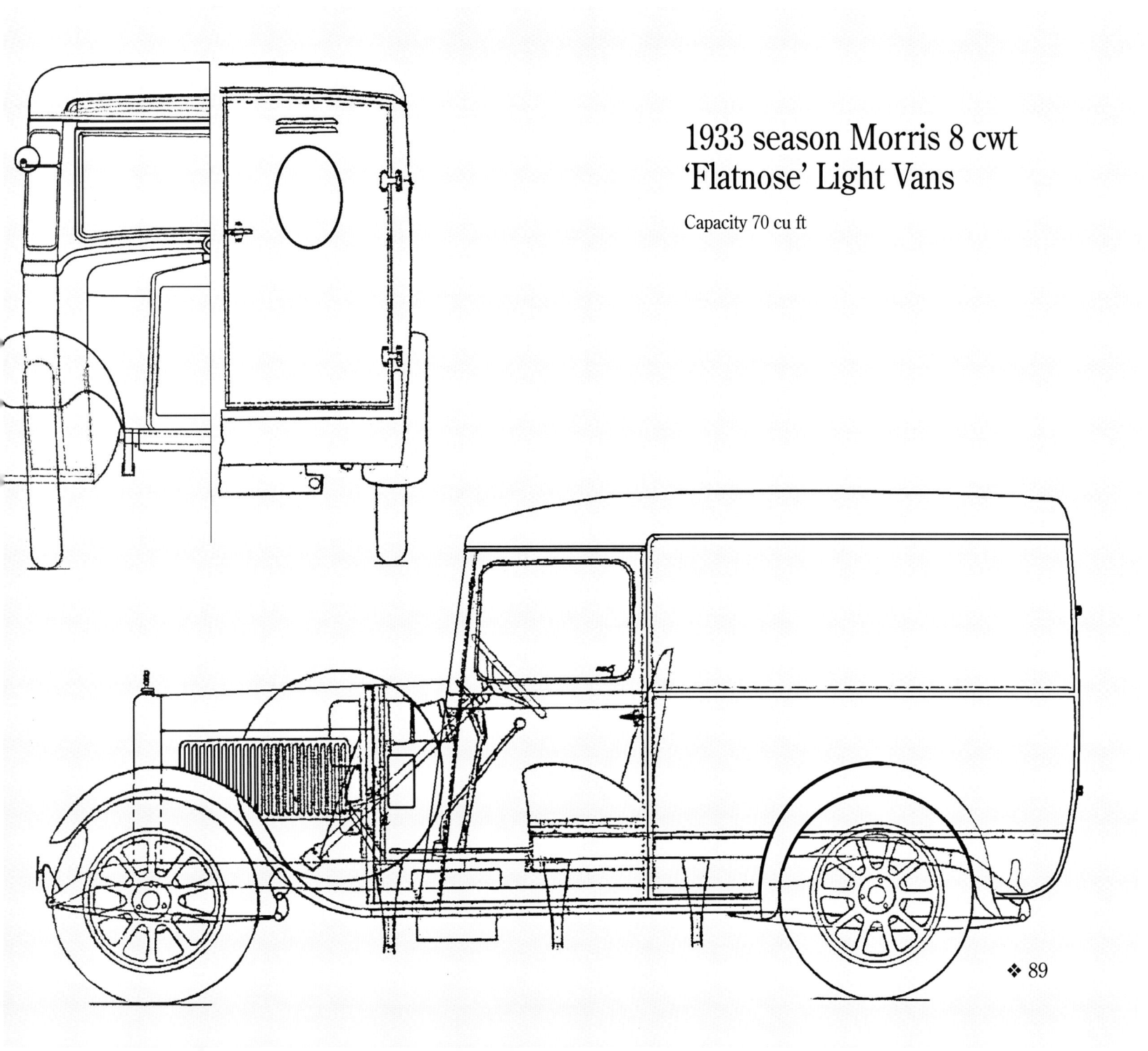

Capacity 70 cu ft

1933 season Morris 8 cwt 'Flatnose' Light Vans

THE MORRIS LIGHT VAN

● The ever popular Morris Light Van has been vastly improved for the 1933 season. The output of the powerful and highly efficient 14/32 h.p. power unit has been further improved, ensuring rapid acceleration and excellent hill-climbing, even with a full load. It is fitted with the well-proven multi-plate cork insert clutch and 3-speed gearbox. The entirely new and attractive body provides a greater carrying capacity and measures 4 ft. 3½ in. in width, 4 ft. 6 in. in length behind the driver's seat, and 3 ft. 10 in. in height, giving a capacity of 70 cubic feet. The equipment is on a very generous scale, and the finish is in shop grey ready to receive the finishing colour most suited to your requirements.

The Morris Light Van is fitted with Triplex safety glass to the windscreen and windows, and all external bright parts are chromium finished.

1933 season Morris 8 cwt 'Flatnose' Light Vans

1934 season Morris 8/10 cwt Light Vans

Manufactured by Morris Motors Ltd., Cowley, Oxford.

Commenced car no. 382398 (13th Sept. 1933).
Finished at car no. 384517 (13th July 1934). *Note*: Although 1934 season Morris Cowley cars had a prefix to their chassis numbers, to denote year and model – e.g. 34/C 1234 – a prefix was not added to the chassis numbers of Morris 8/10 cwt Light Vans until the 1935 season – e.g. 35/TWV 7721.

Body types/features. Built at Coventry (Bodies Branch). Mounted and finished at Cowley. Body is similar to 1933 season Light Van with redesigned front doors/scuttle/windscreen and modified to suit 1934 season chassis frame and rear mounted fuel tank.

Capacity. 8/10 cwt. 75 cu ft.

Windscreen frame. Integral with body, openable and is raked.

Front wings. Pt nos. 50714 (o/s), 50715 (n/s). A headlamp mounting bar is fitted to the front wings and passes in front of the radiator.

Rear wings. Pt nos. 3637 (n/s), 3638 (o/s). Common to 1933 season Vans and 1929 season Morris Cowley 2 and 4 seaters.

Chassis frame. Pt no. 50629 – home market. Pt no. 51807 – export. Cross tube, third and rear cross members common to 1932 season Morris Cowley car.

Engine. Type CS. 13.9 h.p. (*see page 40*) Chain driven camshaft. 'Turbulent' type (Ricardo) cylinder head. Coil ignition with distributor driven from camshaft. Water pump fitted to export models. Car type 'fume consumer' *not* fitted. Twin clutch plates fitted with 36 corks each.

Fuel system. SU type HV2 carburetter. Initially fitted with an SU Petrolift to pump fuel from tank, mounted on the chassis behind the rear axle. Petrolift replaced by an SU petrol pump from car no. 383721.

Transmission. 3 speed gearbox. Similar to 1933 season with laygear supported by caged needle roller bearings. Gear lever spring-loaded. Prop. shaft enclosed by torque tube.

Steering box. Bishop Cam.

Brakes. *Foot* – hydraulically operated. *Hand* – cable operated.

Road springs. *Front* pt no. 2694 – common to early 1931 season vans and late 1928 & 1929/30 cars and vans. *Rear* pt no. 2155 – as fitted from 1927 season vans. (Armstrong shock absorbers fitted all round).

Rear axle ratio. 5.27:1.

Radiator. Pt no. 51090 (*see page 47*).

Price. With body in 'shop grey' £160.0.0
      with body painted £163.10.0
      chassis £125.0.0

Unladen weight. 19 cwt. 2 qtrs. 26 lbs.

Wheelbase. 105".

Track. $48\frac{1}{2}$".

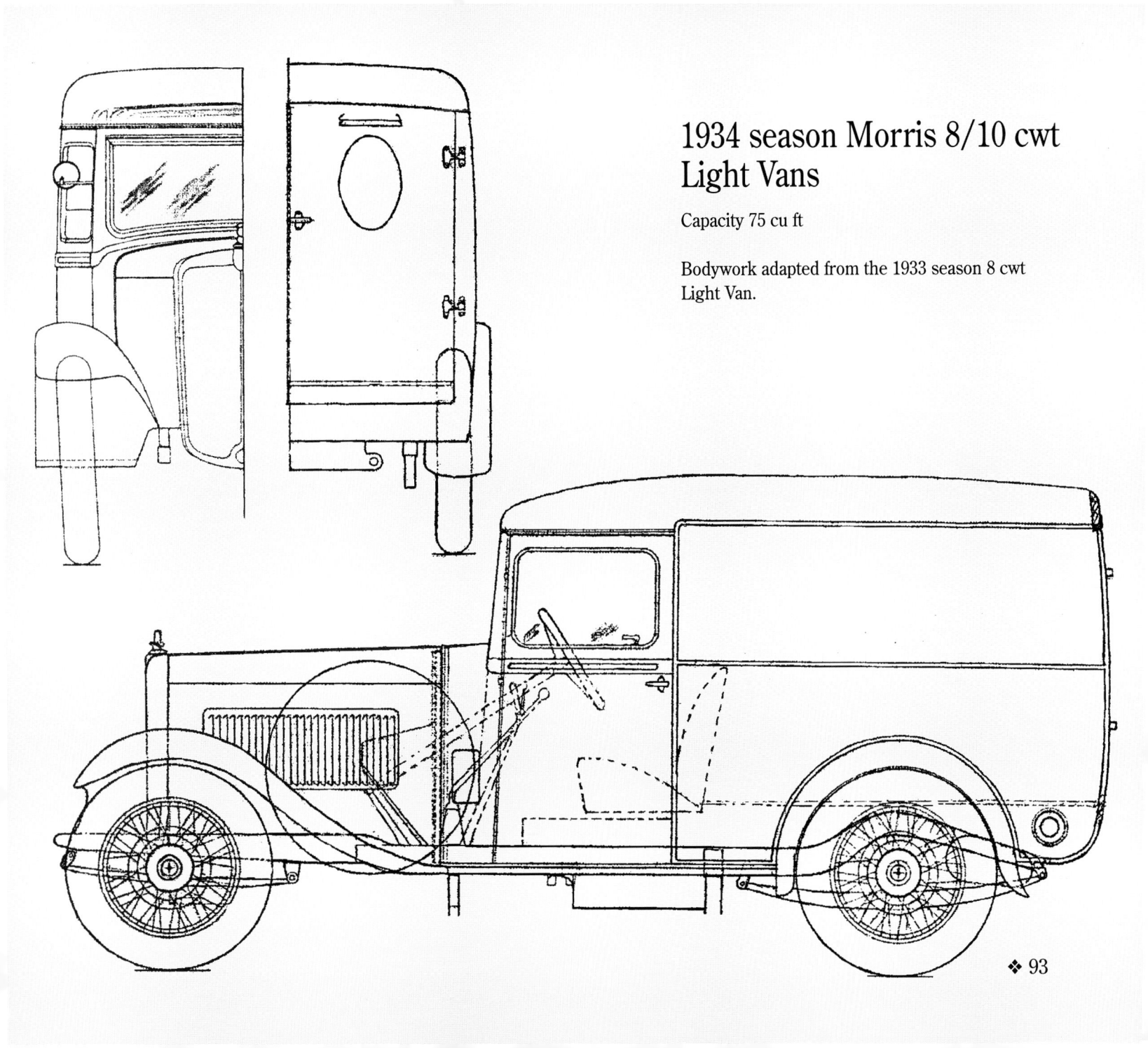

1934 season Morris 8/10 cwt Light Vans

Capacity 75 cu ft

Bodywork adapted from the 1933 season 8 cwt Light Van.

1934 season Morris 8/10 cwt Light Vans

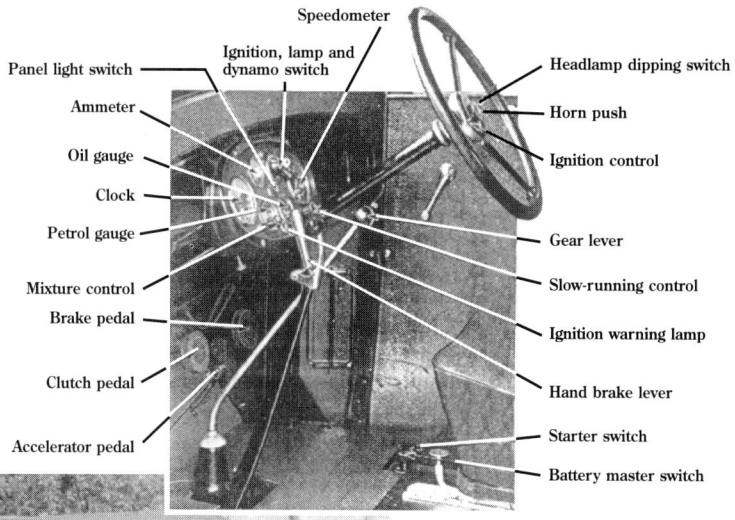

The controls of a 1934 season Morris 8/10 cwt van

1934 season Morris 8/10 cwt Light Vans

STILL MORE MORRIS LIGHT VAN ADVANTAGES

POINTS OF THE 1934 8-10 CWT. MODEL

- 14/32 h.p. side-valve power unit in new design frame with lower centre of gravity.
- Lockheed hydraulic 4-wheel brakes giving much greater stopping power.
- New pattern radiator with stoneguard, improved body with still smarter lines. Capacity 75 cubic feet, *plus* room beside driver.
- Choice of standard colour schemes: Blue, Black, Brown or Green - - **£163 10s.**

Shop Grey **£160 0s.**

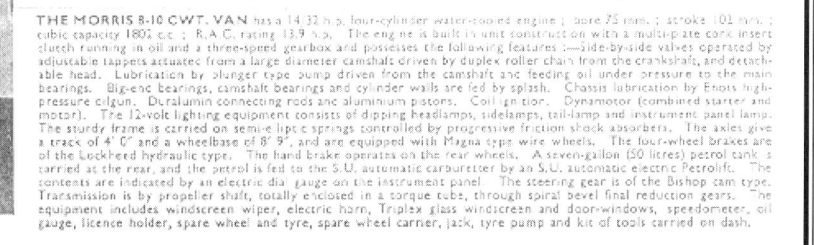

THE MORRIS 8-10 CWT. VAN has a 14/32 h.p. four-cylinder water-cooled engine; bore 75 mm.; stroke 101 mm.; cubic capacity 1802 c.c.; R.A.C. rating 13.9 h.p. The engine is built in unit construction with a multi-plate cork insert clutch running in oil and a three-speed gearbox and possesses the following features :—side-by-side valves operated by adjustable tappets actuated from a large diameter camshaft driven by duplex roller chain from the crankshaft, and detachable head. Lubrication by plunger type pump driven from the camshaft and feeding oil under pressure to the main bearings. Big-end bearings, camshaft bearings and cylinder walls are fed by splash. Chassis lubrication by Enots high-pressure oilgun. Duralumin connecting rods and aluminium pistons. Coil ignition. Dynamotor (combined starter and motor). The 12-volt lighting equipment consists of dipping headlamps, sidelamps, tail-lamp and instrument panel lamp. The sturdy frame is carried on semi-elliptic springs controlled by progressive friction shock absorbers. The axles give a track of 4' 0" and a wheelbase of 8' 9", and are equipped with Magna type wire wheels. The four-wheel brakes are of the Lockheed hydraulic type. The hand brake operates on the rear wheels. A seven-gallon (30 litres) petrol tank is carried at the rear, and the petrol is fed to the S.U. automatic carburetter by an S.U. automatic electric Petrolift. The contents are indicated by an electric dial gauge on the instrument panel. The steering gear is of the Bishop cam type. Transmission is by propeller shaft, totally enclosed in a torque tube, through spiral bevel final reduction gears. The equipment includes windscreen wiper, electric horn, Triplex glass windscreen and door-windows, speedometer, oil gauge, licence holder, spare wheel and tyre, spare wheel carrier, jack, tyre pump and kit of tools carried on dash.

10

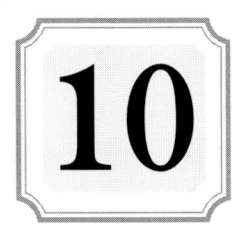

E. G. Wrigley & Co. Ltd. and Morris Commercial Cars Ltd., Birmingham 1898 to 1936

The firm of E. G. Wrigley & Co. Ltd., was registered on 18th May 1898, to acquire the tool making business set up by Edward Greenwood Wrigley at 232 Aston Road, Birimingham. In 1902, the company moved to new purpose-built premises in Foundry Lane, Soho, Birmingham and, in addition to being tool makers, they soon become specialist manufacturers of gearboxes, axles and steering gear for several makes of vehicle, with a reputation for good quality. Rapid expansion took place during the First World War when the factory switched to war work, which included the manufacture of aircraft parts and gearboxes for tanks. By 1920, the Chairman of the company was J. A. Kendrick with J. D. Stevens as Managing Director and F. G. Woollard as Assistant Managing Director.

Frank Woollard had joined Wrigleys in 1911 as Chief Designer and soon after his appointment he began collaborating with W. R. Morris over the design and manufacture of axles and steering sets for the first 'White & Poppe' Morris Oxfords. Woollard, who was destined to have a distinguished career within the motor industry, later joined Morris Engines Ltd., Coventry, as detailed in *Chapter 5*.

The sharp rise in the demand for new cars after the end of the First World War encouraged Wrigleys to enter into an agreement, during 1919, with Sir William Angus Sanderson & Co. and Messrs. J. Tylor & Sons Ltd., with the idea of pooling their skills to mass produce vehicles. The arrangement was that Tylor would manufacture the engines (14.3 h.p., 4 cylinder, side valve with a capacity of 2305cc), Wrigleys would make the gearboxes, front and rear axles and torque tubes, whilst Angus Sanderson

Right: This advertisement appeared in October 1923, shortly before E. G. Wrigley & Co., Ltd. went into receivership. The 1244cc engine shown, which has a single overhead camshaft driven via a vertically mounted dynamo, could well have influenced the design of the 847cc Morris Minor/M.G. 'M' Type Midget engine that was manufactured by Wolseley Motors Ltd. from 1928 until 1932

would make the bodies and undertake the final assembly, distribution and selling of vehicles. The agreement also permitted a director from each company to have a seat on the board of the other companies.

Because they were so confident about the volume of business that would be generated, Wrigleys advertised that they could not accept any more orders from other manufacturers and, at the same time, decided that extra capital was required to provide the facilities to produce the projected quantities of components. In 1919, Wrigleys authorised capital stood at £200,000 and during 1920, 1,300,000 £1 Ordinary shares were issued, thereby raising the authorised capital to £1,500,000. It was a bold move which was to have disastrous consequences for both the company and its shareholders, although some employees were to be more fortunate as they eventually found themselves working for Morris Commercial Cars Ltd.

It had been hoped that the 14hp Angus Sanderson would compete successfully with the Model 'T' Ford but it never lived up to its promise and only about 3000 were built.* In any case, by the time the car was announced, the post war boom had started to collapse so the expected demand for Wrigleys products did not materialise and they started to get into financial difficulties.

Left: A 1925 Morris-Commercial 12 cwt 'L' type Van

* Sir William Angus Sanderson & Co. went into receivership in 1921 and was sold to G. E. Ostwalt, Managing Director of British Spyker Co., Ltd.

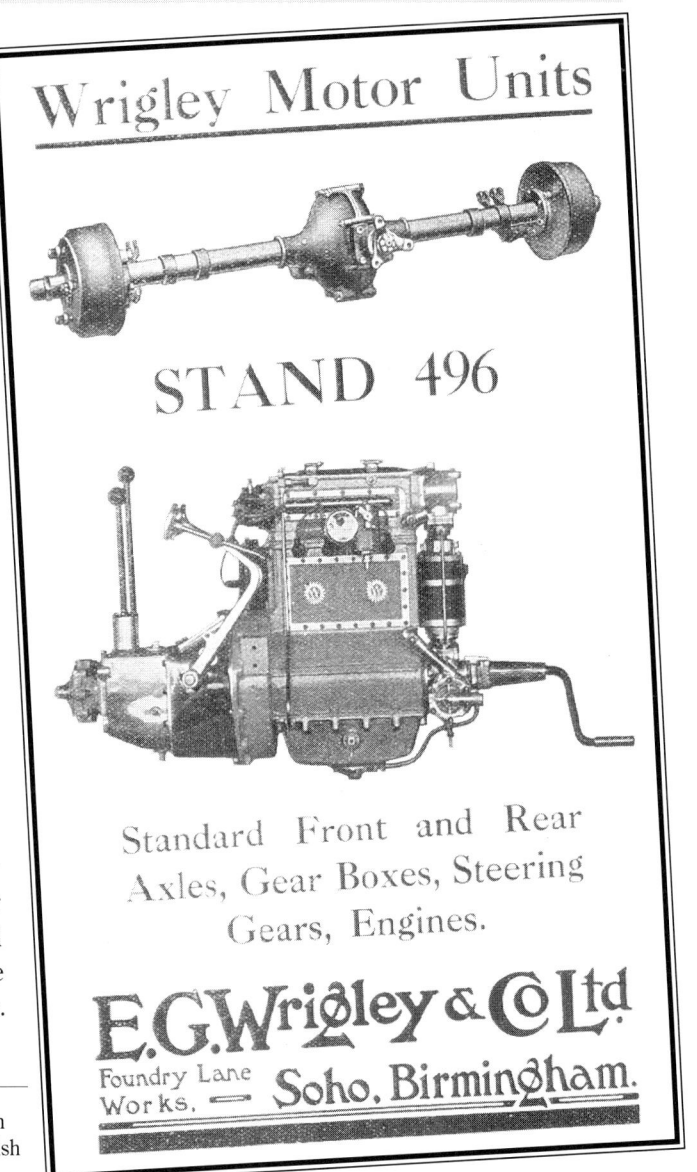

Wrigley Motor Units

STAND 496

Standard Front and Rear Axles, Gear Boxes, Steering Gears, Engines.

E.G. Wrigley & Co Ltd

Foundry Lane Works, Soho, Birmingham.

❖ 97

At the annual general meeting held in October 1920, Wrigley's chairman announced that for the year ending 30th April 1920, the company had shown a loss of £144,226, owed Lloyds Bank £280,804 and its interim dividend of £20,000 had absorbed the general fund. The shareholders then demanded the meeting to be adjourned for one month, so that in the meantime they could have private consultations with the directors. The situation deteriorated further until, in February 1922, Wrigley's authorised capital was reduced, by order of the High Court, from £1,500,000 to £1,106,307 by writing 12/- per share off the 656,155 issued Ordinary shares. Although Wrigleys continued to struggle to maintain its business, the company was eventually put up for sale in 1923, but the receivers were called in during December of that year because no buyer could be found.

Another insight into Wrigley's demise can be found in a paper presented to the Insitute of Automobile Engineers by Cecil Kimber* during 1934. Kimber, who had been working for Wrigleys as a 'Works Organiser' when they got into financial difficulties, well remembered his horror and amazement when one day he discovered that the managing director of the concern was totally unable to read an ordinary trading account, and could not tell the debit side from the credit. Being a shareholder of Wrigleys, Kimber lost practically all his hard earned savings, but in 1921 his fortunes turned when he left the company and joined The Morris Garages as Sales Manager.

Above: C. Kimber

---

* Cecil Kimber was educated at Southport Grammar School and then at the Manchester Technical School where he studied accountancy at evening classes. He undertook an apprenticeship at his father's printing machinery business and then worked for Sheffield Simplex, A.C. Cars and Martinsyde Aircraft before joining E. G. Wrigley & Co. Ltd., of Birmingham as an 'organisation expert'. When Wrigleys got into financial difficulties during 1921, Kimber left the company and joined The Morris Garages of Queens Street, Oxford, as Sales Manager. In March 1922, Kimber became the General Manager of The Morris Garages, when Edward Armstead, the previous incumbent, resigned. (Armstead. who committed suicide a few weeks after resigning, had purchased W. R. Morris's cycle business in 1908 and had been General Manager of The Morris Garages since 1913). During the mid 1920s, The Morris Garages marketed Morris Oxford and Cowley chassis ('Bullnose' and 'Flatnose') with special bodies and the sporting variants were initially advertised as 'M.G Super Sports Morris' thereby founding the 'MG' marque. (The Morris Garages remained in W. R. Morris's personal ownership until 1962 when he gave the company to the Nuffield Foundation). From 1928, the cars produced by The Morris Garages were sold as 'M.G.s', a marque distinct from 'Morris'. On 21st July 1930, the M.G. Car Company Ltd. was registered, with W. R. Morris as Governing Director, 'to take over the manufacturing side of The Morris Garages Ltd.', and with C. Kimber, W. Hobbs and A. Walsh as directors. On the 13th August 1931, Kimber was given official permission to use the title 'Managing Director' of the M.G. Car Company Ltd. but, when the company was sold to Morris Motors Ltd. on 1st July 1935, this title was given to L. P. Lord and Kimber was then given the title of Director and General Manager. On the same day (i.e. 1st July 1935), Kimber was also made a director of Morris Motors Ltd. (As Morris Motors Ltd. was then the largest motor manufacturer in Europe, Kimber then held a more senior position than previously so apparently he did not suffer a loss of favour at this time as has sometimes been suggested.) Kimber was re-appointed as Managing Director of the M.G. Car Co., Ltd. on 24th August 1936, after Leonard Lord resigned. Following a dispute with Miles Thomas (*see Profile on page 115*) in November 1941, Kimber left the Nuffield Organisation and joined Charlesworth Motor Bodies, but, within a year, resigned to take up the position of Works Director with the Specialloid Pistons Co. Kimber was killed in a railway accident at Kings Cross Station, London on Sunday 4th February 1945, a few weeks before his 57th birthday.

W. Hobbs

This advertisement appeared in August 1924

SHIPSIDES
—MAIN DISTRIBUTORS FOR—
Morris Cars, Vans & Lorries

MORRIS ONE TONNER

MORRIS ONE TON TRUCK — £225

Chassis	£185
With Lorry Body	£225
With Tipping Body	£235
With Van Body	£235
Van De Luxe	£250

MORRIS 8 CWT. VAN

MORRIS VAN DE LUXE — £200

Standard Van	£180
De Luxe Van	£200
Commercial Traveller's Car	£190

14-seater Charabanc (Convertible for passengers and goods) £345

MORRIS-COWLEY

2-Seater	£175
Occasional Four	£185
4-Seater	£195

MORRIS-COWLEY 4-SEATER — £195

MORRIS-OXFORD

	With FRONT BRAKES	Without FRONT BRAKES
2-Seater	£260	£250
4-Seater	£285	£275
Coupe	£305	£295
Saloon (4 doors)	£385	£375

Buy British and be Proud of it
Early Deliveries of 1925 Models

Exchanges and Deferred Payments arranged on the most favourable terms

DAYBROOK, NOTTINGHAM

He later became a director of both Morris Motors Ltd. and the M.G. Car Co., Ltd.

It so happened that at the time of Wrigley's collapse, W. R. Morris was anxious to expand his business by entering into the manufacture of purpose-built commercial vehicles and the potential offered by Wrigleys as a going concern did not go unnoticed. W. R. Morris, therefore, paid the receivers £213,044 (about £8.5 million at 1999 values) for the business, which he took over on the 1st January 1924. The name on the factory's headboard was changed to 'W. R. Morris, successor to E. G. Wrigley & Co., Ltd.', and this remained until 4th February 1924 when it became Morris Commercial Cars Ltd. It was a shrewd purchase, for not only had W. R. Morris bought factory space in the heart of the industrial Midlands, he had also acquired a large quantity of machines, tools and a skilled workforce that was already familiar with designing and manufacturing components for motor vehicles.

Morris Commercial Cars Ltd. was set up with W. R. Morris as Governing Director, R. W. Thornton* and L. W. Pratt (*see Profile page 114*) as directors and Wilfred Hobbs as Secretary and Chief Accountant. During November 1924, Edgar H. Blake, who had been the Sales Manager of the Dunlop Rubber Co. Ltd., became the Managing Director of Morris Commercial Cars Ltd., and he held this position until July 1926 when Morris appointed him as his Deputy Governing Director at Cowley, after H. W. Grey, the

---

* R. W. Thornton was from Messrs. Thornton and Thornton of Oxford who were W. R. Morris's auditors.

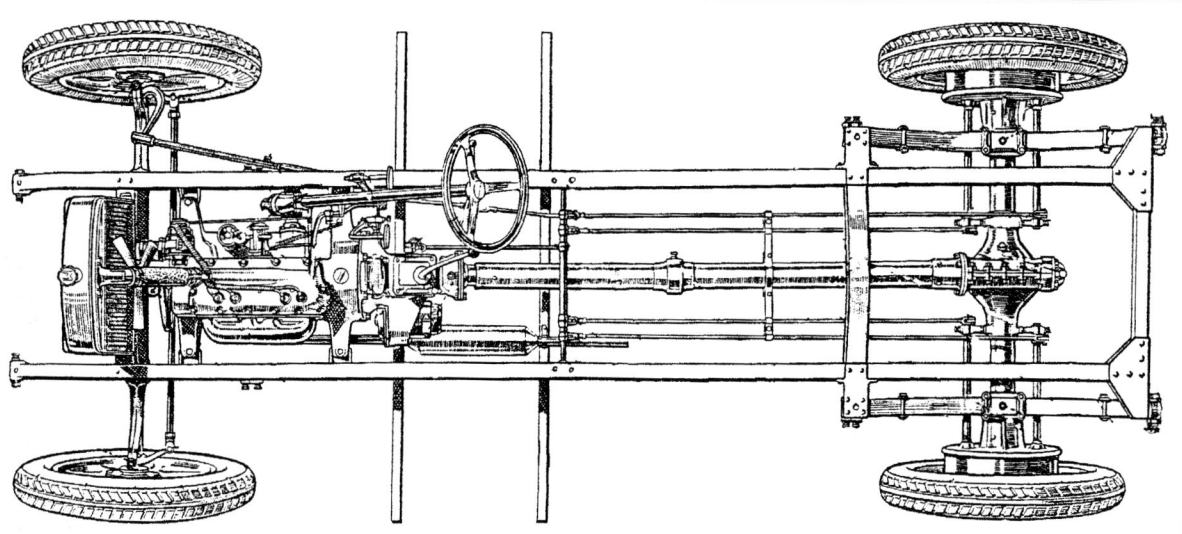

Left: Plan view of a Morris-Commercial 'T' type One-Ton chassis, fitted with a 13.9 h.p. type CE engine

C. F. Lawrence King

W. Cannell

previous incumbent, stood down owing to ill health. The General Manager of Morris Commercial Cars Ltd., William Cannell, then took over as its Managing Director, with C. F. Lawrence-King as his Sales Manager. When Cannell resigned on 8th July 1927, because Morris had given him the task of setting up Wolseley Aero Engines Ltd., William Wilson Hamill was appointed initially as General Manager, and then, on the 15th November 1927, as Managing Director. Hamill remained in this position until February 1932, when the post was taken over by Oliver Boden, who then appointed Miles Thomas as director and General Manager (*see Profile page 115*). During a merger of W. R. Morris's personally owned companies, Morris Commercial Cars Ltd. was sold to Morris Motors Ltd. in October 1936 (*see page 32*).

With 13.9 h.p. type CE engines and gearboxes, which were already being made by Morris Engines Ltd. of Coventry for the 'Bullnose' Morris Oxford, and with front and rear axles and steering mechanisms from the plant and workforce that had been E. G. Wrigley & Co., Ltd., the only major items needed to create the Morris 'T' type One-Ton Truck were the chassis frame/bulkhead, road springs and radiator, which, once assembled, could be fitted with a cab and truck body. The responsibility for this vehicle lay with Morris Commercial's Chief Designer, P. G. Rose, who held this position until 1937, when he became the Chief Engineer of the company. During the early part of 1927, W. R. Morris presented P. G. Rose and his team at Morris Commercial Cars Ltd. with the task of designing the Morris Minor, which was announced at the Olympia Motor Show in 1928. This design not only formed the basis of the M.G. 'M' Type Midget but also the Morris Minor van, which was purchased in large numbers by the GPO.

Between 1932 and 1940 the GPO purchased over 7000 Morris Minor 35 cu ft vans for their Royal Mail and Post Office Telephone fleets and two examples are shown here. The 35 cu ft Royal Mail van (below) was registered AYP 326 in May 1934, whilst the van on the left, registered GW 2426 (chassis no. SV 9748), was one of the first batch of six Morris Minor vans to be introduced into the Post Office Telephones fleet in Feb. 1932 for trial. The glass observation panels above the raked windscreen were provided so that the driver could inspect overhead telephone lines from within the vehicle; bodywork design being the result of collaboration between W. Harold Percy, Ltd., of North Finchley, London, and Post Office engineers. The Morris Minor car, from which these vans were derived, was designed by Percy Rose, Chief Designer of Morris Commercial Cars Ltd.

P. G. Rose

The first Morris-Commercial 'T' type One-Ton truck (chassis no. 001) being shown to Gordon Stewart (with waistcoat and chain) of Stewart & Ardern, Morris main dealers for the London area, by W. R. Morris (standing nearest to camera) outside the building in Hollow Way, Cowley that housed W. R. Morris's office. This vehicle, which was driven down to Cowley on 'trade plates', was registered HA 2064 on 25th May 1924 and during its journey from Birmingham, it averaged 24 miles per gallon of petrol. Note the fuel consumption test tank mounted on the bulkhead and that the vehicle is badged as a 'Morris'. Initially, Morris 8 cwt 'Snubnose' vans had been badged as 'Morris-Commercial', but when they were renamed 'Morris Light Van' at the beginning of the 1927 season, the products of Morris Commercial Cars Ltd. were then badged as 'Morris-Commercial'. (See pages 48 & 112). The vehicle is still in existence, having been restored by apprentices at Morris Commercial Cars Ltd. during the early 1960s

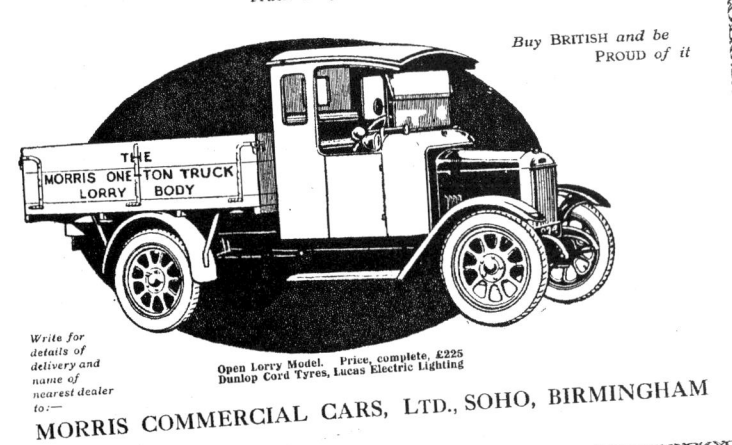

This advertisement, which appeared in October 1924, alludes to the Model 'T' Ford's jocular image as the 'Tin Lizzie', and that a British ton is heavier than a U.S. ton.

Percy George Rose had completed an apprenticeship with Royce & Co. (the forerunners of Rolls-Royce) prior to the First World War and had joined Morris Motors Ltd. at Cowley in 1922, when the manufacture of a one ton truck was first mooted. Much preliminary work was needed to determine the feasibility of this project and P. G. Rose was selected to carry out the necessary investigations, as well as preparing the original design of the vehicle, apparently in great secrecy. Soon after W. R. Morris bought E. G. Wrigleys & Co., Ltd., P. G. Rose relocated to Birmingham and became involved in planning and reorganisating the Soho factory for commercial vehicle production.

The final decision to manufacture the Morris One-Ton Truck, now popularly known as the 'Tonner', was taken by W. R. Morris towards the middle of 1923, and his intention was announced to the public in a detailed article that appeared on 28th January 1924 in the *Motor Transport* magazine. This was only four weeks after W. R. Morris had acquired Wrigley's factory where the vehicle was to be made.

During the early 1920s, while the heavier side of the transport industry was fairly well established, the use of light commercial vehicles in large numbers was just beginning, owing mainly to the availability of pneumatic tyres with cord construction. These tyres, which had been introduced into Britain from the USA at the end of 1921, solved the problems and frequent failures associated with those tyres using canvas. By 1923, the one ton (payload) truck was rapidly becoming popular in the UK and the demand for them was being satisfied principally by American products. At this time, British manufacturers of heavy commercial vehicles seemingly paid little heed to the demand for lighter models, and this situation gave W. R. Morris an added incentive to manufacture them.

Until cord tyres became available, commercial vehicles had had to be fitted with solid rubber tyres, as tyres of canvas construction were unsuitable, so

A view of the 'assembly line' at Morris Commercial Cars Ltd., Soho, Birmingham, during 1924. On the left is shown the wheel hoist and runway. In the foreground the springs and front axles are being added to the upturned frames and in the background are the six phases of the chassis assembly. The chassis leaves the line with its engine running and at once commences a road test

Above: An aerial view of Morris Commercial Cars Ltd., Foundry Lane, Soho, Birmingham, c.1928. W. R. Morris bought this factory, complete with machines, tools and skilled workforce, on 1st January 1924 from the liquidators of E. G. Wrigley & Co., Ltd.

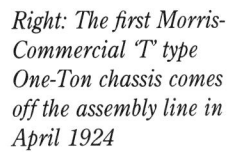

Right: The first Morris-Commercial 'T' type One-Ton chassis comes off the assembly line in April 1924

❖ 105

MORRIS-COMMERCIAL 12-cwt. Van

Popular model
Price complete, ready for the road
£175
Complete with Spare Wheel and Tyre
At Works

THE MORRIS-COMMERCIAL 12-cwt. VAN is designed to meet the requirements of transport users demanding an exceptionally sturdy vehicle with large capacity body for speedy all-the-year-round delivery. Though robustly built, appearance has also been considered, and it provides a valuable advertisement for the owner's business when suitably painted.

The price is highly competitive, and, as running costs and maintenance are extremely light, it forms an excellent business proposition.

Large numbers of these vehicles are in regular use by the General Post Office and many important stores, manufacturers, etc.

Buy this Van out of your Profits!

The Morris-Commercial 12-cwt. Van may be obtained upon a most attractive hire-purchase basis financed by Morris Commercial Cars Ltd. This enables the owner to buy out of income, or with the profits which his vehicle makes, thus leaving his capital free for other uses.

	Deposit	12 monthly payments of
Popular Van	£43 15 0	£11 11 11
De Luxe Van	£49 10 0	£13 2 4

12-cwt. Popular Van

The general design of this van lends itself admirably to numerous trades requiring a speedy delivery service which can be relied on all the year round, under the most strenuous conditions. The underframe and framework is of selected ash, panelled in aluminium sheet, ensuring freedom from rust. The roof is covered with plywood, over which is stretched good quality waterproof canvas. Half-door fitted to driver's cab. The two rear doors, over the full width of van, give full access to a roomy body. Oval windows are embodied in the rear doors, giving a clear vision for the driver. The driver's seat and back-rest are fitted in a comfortable position. Leading dimensions detailed below.

	Length	Width	Height
Overall dimensions	13 ft. 6 in.	5 ft. 8½ in.	7 ft. 2 in.
Inside dimensions	5 ft. 2 in.	4 ft. 4 in.	4 ft. 4 in.

Loading height—2 ft. 8½ in.
Length behind driver's seat—5 ft. 2 in.

12-cwt. De Luxe Van

As in the case of the Popular Van, the De Luxe Van is robustly built to withstand rough, hard usage and the severest test of day-in and day-out service. This van is specially built and artistically moulded, having a coach door with drop window and comfortable driving position. The underframe and framework is of selected ash. Panelled in aluminium-armoured plywood. The roof is covered with plywood, over which is stretched waterproof canvas. Two rear doors open to the full width of the van. The floor is built with wheelarches to give a low loading line. All doors are fitted with good quality locks, hinges, etc. This vehicle is highly suitable for light, speedy conveyance of goods. Leading dimensions are detailed below.

	Length	Width	Height
Overall dimensions	13 ft. 6 in.	5 ft. 8½ in.	7 ft. 2 in.
Inside dimensions	5 ft. 2 in.	4 ft. 4 in.	4 ft. 4 in.

Loading height—2 ft. 8½ in.
Length behind driver's seat—5 ft. 2 in.

SPECIAL NOTE
The vehicles quoted in this list are delivered in a coat of priming colour only. Prices for painting in choice of colours furnished upon application.

De Luxe model complete £198
At Works

The large capacity Van that's cheap to run

This brochure for the 'L' type 12 cwt van was published in April 1929

Right: This advertisement appeared in September 1927

Above: The rear axle assembly of a Morris-Commercial 'T' type 'Tonner'. The three-quarter floating type axle has an overhead worm final drive and 15" brake drums

Below: Morris Commercial Cars Ltd., offered a range of Standard model public service bus bodies on the 'T' type One-Ton chassis.

BRITAIN'S ANSWER!

The New
MORRIS-COMMERCIAL
Light-Tonner

The Greatest Value-for-Money Truck ever built.

Glance over these unique features:

22-25 M.P.G. Petrol.   900-1000 M.P.G. Oil.
12,000-15,000 Miles per set of Tyres.
Engine-driven Tyre Inflator, Speedometer, Electric Horn.

Subject to only £16 Tax.

HIRE-PURCHASE—You can obtain this remarkable new Truck for £51.5.0 down and 12 monthly payments of £13.11.8, or in the case of the Van, a first payment of £53.15.0 and 12 monthly payments of £14.4.10.

NOTE—This Light-Tonner is an entirely new model and should not be confused with the standard Morris-Commercial heavy duty Ton Truck, which remains at £225 complete.

May we send you our attractively illustrated Brochure No. B4 which gives full particulars of this new model?

PAYS ONLY £16 TAX

£205 TRUCK
£215 VAN

MORRIS-COMMERCIAL

MORRIS-COMMERCIAL CARS LTD., SOHO, BIRMINGHAM

their speed was restricted and, as a result, these vehicles were often slower and less economical than other forms of transport. The introduction of pneumatic cord tyres dramatically changed the prospects for commercial vehicles and suddenly transformed them into more economical forms of transport. Nevertheless, manufacturers initially faced difficulties in persuading customers about the benefits of cord tyres, as pneumatic tyres generally were thought to be unreliable. Consequently, the Morris brochure introducing the 'Tonner' quoted the following:

> 'We desire all purchasers of Morris One-Ton Trucks to note that the guarantee becomes null and void if the pneumatic tyre equipment fitted as standard be replaced by solid or semi-solid tyres (either on all wheels or on rear wheels alone). The Morris One-Ton Truck has been designed to run on pnuematics, and the 32"x4½" DUNLOP STRAIGHTSIDE CORD TYRES provided as standard are of such a size and so robustly made, that tyre trouble is practically eliminated'.

The first Morris-Commercial 'T' type One-Ton chassis left the assembly line in April 1924 and soon afterwards, the vehicle, complete with cab and truck body, was driven down to Cowley for W. R. Morris's inspection. By the end of 1924, 2,486 'Tonners' had been assembled and production more than doubled in 1925, when an average of just over 100 per week were being made, to give a total for the year of 5,399*. This level of production is a noteworthy achievement, bearing in mind that the 'Tonner' was Morris Commercial's

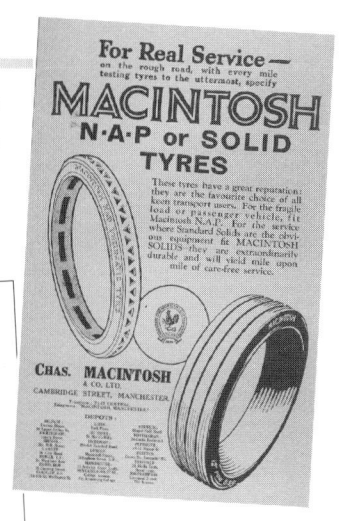

---

* To put this output into perspective, the total production of the M.G. 'M Type' Midget (M.G.'s best selling pre-war model) between 1929 and 1932, was 3,235.

Left: The introduction of cord tyres to the U.K. in 1921 dramatically changed the prospects for commercial vehicles and suddenly turned them into more economical means of transport

Right: Introduced in October 1930, this 100 cu ft Morris-Commercial 'L2' 10 cwt 'De Luxe' van was priced at £190. A 'Popular' model with a 3 lamp set, a 6 volt dynamo in lieu of a 12 volt dynamotor and less a mechanically-driven tyre pump (fitted as standard on 'L2's at this time), became available from mid-1931 at £170. These vans were advertised as 'the all-British van for fragile loads' and having 'springs made from silico-manganese steel giving extraordinarily smooth travel over the roughest roads', but they apparently competed unsuccessfully with the 1931 season 78 cu ft Morris 'Flatnose' 8 cwt Light Vans (see pages 80 & 81), which were priced at £160. [Note that Morris 'Snubnose' Light Vans made between 1927 and 1930 were designated as 10 cwt]. Being derived from the Morris-Commercial 'L2' 15 cwt van, the chassis of the two wheel brake 'L2' 10 cwt van was more robust than that of the 'Flatnose' Light Van, which had four wheel brakes, although both vans had similar 13.9 h.p. engines and 3 speed gearboxes. As there was little demand for Morris-Commercial 'L2' 10 cwt vans, the model type was deleted in 1932.

Below: An aerial view of Morris Commercial Cars Ltd., Adderley Park, Birmingham. This factory was purchased by W. R. Morris during February 1927 from the liquidators of Wolseley Motors Ltd.

first product and that a competitor, the Ford 'One Ton Truck', was over 40% cheaper. (Subsequent events soon gave Morris Commercial Cars Ltd. an advantage over their main competitors, as already discussed, [see pages 15 and 16]).

During September 1925, Morris Commercial Cars Ltd. introduced the 12 cwt 'L' type which was derived from the 'T' type but had a shorter wheelbase and a smaller type CB 11.9 h.p. engine, as fitted to the 'Bullnose' Morris Cowley. It also had lighter axles and although its transmission was initially enclosed by a torque tube as on the 'Tonner', later 'L' types had an open prop shaft with fibre couplings at each end. Two years after the introduction of the 'L' type, the 'LT', or 'Light Tonner', appeared. The 'LT' was basically an 'L' type with a type CE 13.9 h.p. engine and larger brakes.

In 1930, the 'T' and 'L' types were developed into the 'T2' and 'L2', whilst the 'LT' type was discontinued. Unlike its predecessor, the 'L2' was offered with a 13.9 h.p. engine and payload capacities of either 10 cwt or 15 cwt (the 10 cwt model was deleted in 1932). Yet another development took place in 1938 when the 'T3' and 'L3' were introduced. Even though the external appearance changed with each development of these models, their mechanical components remained similar, although their specification was updated with such devices as four wheel brakes, the operation of which was to progress from mechanical to hydraulic, and 4 speed gearboxes.

The development of this line of vehicles has significance to one of the subjects of this book, because large numbers of 'L' 'L2' and 'L3' chassis were supplied to the GPO for its 105 cu ft bodies. Also, short chassis versions of the 'L2' and 'L3', known as the 'L2/8' and 'L3/8', were designed and manufactured specially to suit the GPO's 70 cu ft body. In addition, the GPO operated many Morris-Commercial 'T', 'T2' and 'T3' types within the Royal Mail fleet, and these chassis were usually fitted with 160 and 250 cu ft bodies. The first Morris-Commercial 'T' type One-Ton Royal Mail van, registered XX 1257, was supplied to the GPO during March 1925, less than 12 months after the 'T' type had been introduced. The rapid expansion of the product range meant that from 1927

Morris Commercial Cars Ltd. were also able to supply chassis to suit the GPO's 340 cu ft bodies. At the end of 1931, the company was able to advertise that they could offer vehicles for every trade and purpose from 10 cwt to 5 ton payload capacity, as well as double-decker buses and taxis.

During 1924, 1925 and the early part of 1926, the products of Morris Commercial Cars Ltd. were badged as 'Morris' (*see page 112*), but from the start of the 1927 season (August 1926) they were badged as 'Morris-Commercial'. At the same time, Morris

Above: A Morris-Commercial 'L3', 15 cwt van

Below: This Morris-Commercial 'T' type 'Tonner' was supplied to the GPO in 1928

Morris Commercial Cars Ltd. offered a range of bodywork for the 'T' type One-Ton chassis. The two examples shown here are the Standard Van (top) and De luxe Van (below), which were advertised in 1924 at £235 and £250 respectively

This advertisement appeared in December 1931

8/10 cwt 'Snubnose' vans, which had previously been badged as 'Morris-Commercial', were then badged as 'Morris'. The switch was for corporate and legal reasons, to conform with the laws concerning trade marks, as Morris Motors Ltd. had become a public company during June 1926 (*see pages 31 & 48*). On the other hand, Morris Commercial Cars Ltd. remained the personal property of W. R. Morris until October 1936, when it was purchased by Morris Motors Ltd. The switch also resolved the confusion over which company had manufactured which models.

From February to May 1924, Morris Commercial Cars Ltd. suffered a loss of £22,000, which was hardly surprising as vehicle production had only commenced in April of that year. Consequently, W. R. Morris had to put more money into the business to provide additional working capital and to finance stocks, while sales were building up, but by May 1925 a profit of £39,000 had been made.

The rapidly increasing level of production at the Soho factory, which was boosted by the introduction of import duties on commercial vehicles on 1st May 1926, led to a requirement for more space. The floor area of the factory on 1st January 1924, when W. R. Morris had bought Wrigley's premises, was 130,148 sq ft but by July 1926 it totalled 258,325 sq ft – an extra 98%. A new test and body shop had been constructed together with a service department and additional storage space.

In order to provide yet more space, Morris Commercial Cars Ltd. acquired a factory and land at Adderley Park, Birmingham, from Wolseley Motors (1927) Ltd. on 29th November 1929. The Adderley Park site had been part of the assets of Wolseley

Above: A Morris-Commercial 'T' type One-Ton van fitted with a radiator having an extended header tank to improve cooling

Left: The radiator of a Morris-Commercial 'T' type One-Ton truck, which were badged as 'Morris' until the end of the 1926 season. The legs of the letters 'R' on early badges are shorter as shown on page 102

Above: The radiator badge changed from 'Morris' to 'Morris-Commercial' at the start of the 1927 season

Motors Ltd., which W. R. Morris had purchased from the liquidators during February 1927. (This transfer of assets between two companies owned personally by W. R. Morris exemplifies the purpose of Morris Industries Ltd., as already discussed in *Chapter 4*). The manufacture of Morris-Commercials then commenced at Adderley Park in addition to Soho. Vehicle production at Soho ceased in 1932 and the factory was then used for storage until the premises were sold in 1968.

By the end of the 1930s, Morris Commercial Cars Ltd. had become the largest manufacturer of commercial vehicles in Europe and it remained in a leading position until the closure of the Adderley Park factory in 1974.

This advertisement for the Morris-Commercial L3 and T3 appeared in August 1938

Below: A Morris-Commercial Imperial double-decker operated by Birmingham Corporation

❖ 113

Profiles

L. W. Pratt – 1880 to 1924

L. W. Pratt was the owner of Hollick & Pratt Ltd., bodybuilders of Coventry, until W. R. Morris took over this company in December 1922 (*see Chapter 7*). After the take over, Pratt continued to control Hollick & Pratt Ltd., which became Morris Motors Bodies Branch in 1926, as well as the Morris-owned bodyshop at Cowley. In addition, Pratt became W. R. Morris's second-in-command with the title of Deputy Managing Director and the pair became close friends on both personal and business levels. Morris relied heavily on Pratt for advice and support and the trust that Morris put in Pratt was exceptional. Inevitably, Pratt was involved in Morris's decisions to enter into volume production of car-derived vans and purpose built commercial vehicles. Sadly, Lancelot Pratt died during 1924 at the age of 44, an event that deeply affected W. R. Morris. The following obituary, written by W. R. Morris, appeared in May 1924.

L. W. Pratt

'There come times in a man's career when he sits down and wonders whether anything on this earth is, after all, worth while; times when the whole substance seems to fall out of the universe and the term "success in life" becomes a meaningless, tasteless and insipid nothingness. With my whole heart I confess that I passed through such an experience when my greatest friend, Mr. L. W. Pratt – and the best friend a man ever had – passed away on Saturday, April 19th, at Stratford-on-Avon.

There are some men put on this earth of ours to leaven the lump of humanity and to lighten its burden at all times. Such a man was Mr Pratt. To me he was infinitely more than a partner in business. He was a rock of confidence and the finest comrade any man could wish. Always cheerful, open as a book and yet possessed of an amazing aptitude and foresight in business, he did more – far more – than most people realise to make motoring what it is in this country today.

Things that could never have been accomplished single-handed we did together with certainty and without tiresome deliberation. It was he who, by reorganising the Hollick and Pratt body-making works in Coventry – in a way that was unprecedented in this country – made it possible for the difficult production of bodies to keep pace with the more simple increase in chassis output. When I took over that concern in December, 1922, Mr Pratt, as Deputy Governing Director of Morris Motors Ltd., became more than my second self. With a broad, generous and optimistically balanced outlook on life, for him the word "impossible" did not exist. If a job wanted doing he found the right way to do it – and to do it with a smile.

For business was the salt of his life. Not the mere making of money. But the pride of achievement – that satisfaction of doing something useful for his fellow men and women. His work was his glory and among those who came into contact with him he infused a spirit of happy enthusiasm that made the hardest day of worry and toil seem but an episode for humorous reflection. His very personality over-rode difficulties, smoothed the path and changed one's whole outlook on life when things seemed awry.

It was God's will that he should be taken. So be it. But the good that men do must live after them, and I want everybody who is in any way connected with Morris productions to realise to the full that the late Mr Pratt has done far more for them and for me than mere words can convey. There have been very few men of his calibre ever born. May his name live for ever in the annals of those who have helped to make this world of ours a better place. Whatever success Morris Motors achieve in the future will be due to a very great extent to the splendid groundwork done by Mr. Pratt' – *W. R. Morris*

O. Boden – 1887 to 1940

Oliver Boden O.B.E. was a much respected engineer who had commenced his career with Vickers where he organised the manufacture of armaments during the First World War. He was Works Manager of Wolseley Motors Ltd. when W. R. Morris bought the company in February 1927 and continued to hold this position when he was appointed managing director of Morris Commercial Cars Ltd. in 1932. He became Managing Director of Wolseley Motors Ltd. in 1933, when Leonard Lord relinquished this post to become Managing Director of Morris Motors Ltd. at Cowley (*see page 29*), and was appointed Vice Chairman of the Nuffield Organisation and Morris's deputy in 1936, when Leonard Lord resigned (*see page 33*). He was appointed Deputy Controller of the Spitfire factory at Castle Bromwich, Birmingham in 1938 and the Managing Director of Nuffield Mechanisations and Aero Ltd., a company that made tanks, Bofors guns, etc., during the Second World War. Oliver Boden held a directorship of Morris Motors Ltd. from August 1936 until the 6th of March 1940, when he collapsed and died at the age of 53. After his death, W. R. Morris said of Boden that 'he has been described in an obituary notice as my right-hand man, and no cognomen could be more appropriate. . . the service he rendered to me constantly increased in worth and importance. . . His entire distaste for any form of self publicity has resulted in his name being less widely known than would otherwise be the case. . .'

Oliver Boden O.B.E

W. M. W. Thomas – 1897 to 1980

William Miles Webster Thomas D.F.C. (later Sir Miles and then Lord Thomas of Remenham), who was educated at Bromsgrove School and Birmingham University, served as a fighter pilot in the Royal Flying Corps. during the First World War after which he became a motoring journalist. He joined Morris Motors Ltd. in 1924 to 'take over advertising and sales promotion' and to start *The Morris Owner* magazine. Miles Thomas married Hylda Church, W. R. Morris's personal secretary in 1924 and became a director of Morris Motors Ltd. in May 1927, retaining this position on his appointment as Director and General Manager of Morris Commercial Cars Ltd. during 1934. In 1936 he became Managing Director of Wolseley Motors Ltd. and, in 1940, after Oliver Boden had died, was appointed the Vice Chairman of the Nuffield Organisation and Morris's deputy. In addition to these responsibilities, Miles Thomas joined the board of the Colonial Development Corporation during 1947, but, at the end of the same year, he left the Nuffield Organisation, although he remained in contact with W. R. Morris for the rest of his life. Miles Thomas then became the Deputy Chairman of the British Overseas Airways Corporation and fifteen months later, in 1949, was appointed as Chairman. After seven years in this position he resigned to become the Chairman of the British Monsanto Chemical Company. By 1965, when he was aged 68, Miles Thomas was on the board of several companies, including the Dowty Group Ltd. and the Sun Alliance Insurance Ltd., as well as being the Chairman of the National Savings Committee and the Vice-Chairman of the Welsh Economic Council. Despite owning a large farm in Southern Rhodesia (now Zimbabwe), W. M. W. Thomas spent his declining years at Remenham Court, Henley-on-Thames, Oxfordshire.

W. M.W. (Miles) Thomas D.F.C.

11

Royal Mail Vans

Background Information

Immediately after the First World War the GPO experienced difficulties with its transport contractors and two ex-WD Ford vans and a second-hand GWK van were purchased to enable it to undertake trials at Eastbourne and Kingston-upon-Thames. These trials were so successful that 50 new GWK 8 cwt vans were bought during 1920 and a further 106 GWKs and 200 Fords of various capacities were delivered in 1921. Owing to difficulties with the GWK (*see page 126*), the GPO ceased buying this type and those already purchased were soon disposed of. Consequently, the GPO's fleet of Royal Mail vans then became dominated by Model 'T' Fords (*see page 130*).

During the mid 1920s, the GPO came under considerable pressure to buy British vehicles instead of Model 'T' Fords which, although being assembled at Trafford Park, near Manchester, were of American origin. The pressure on the GPO came from several quarters including 'The Colmore Depot' of Birmingham, a Morris main dealer, whose proprietor Mr. E. C. Paskell publicly accused the Postmaster General of avoiding the question when he was asked why the GPO were using foreign motor vans.* Pressure also came from

Mr. E. C. Paskell

---

*Although popular sentiment was encouraged to regard Fords as 'foreign', by the mid 1920s over 90% of every Trafford Park-built Ford was made in the British Isles.

A 1927 Morris 'Snubnose' 70 cu ft Royal Mail van, used in the Northern District Office area. The chassis (no. 179023) of this vehicle was erected on 27th January 1927 to a 'Plain' specification with two wheel brakes. It was converted to four wheel brakes before being despatched from the factory at Cowley. Note the three lamp set (electric), the two oil lamps and the front wheel hub cap which houses an odometer (see page 139, note 1)

Left: A 1935 Morris-Commercial L2 105 cu ft Royal Mail van. Note the oil lamps and the side lights mounted on the windscreen pillars

Morris Motors Ltd. via the widely-circulated *The Morris Owner* magazine, which published the following:

'A great effort is now being made to stabilise British trade and part of the propaganda is the British Industries Fair, which will be held in Birmingham early next year [1926]. The buyers and agents, when they arrive in Birmingham, the centre of the motor industry, will see the nation's post being carried in conveyances associated with America.'

As a result of all this, the GPO commenced further trials during 1924 with two Morris 'Snubnose' Royal Mail vans (*see pages 123, note 4, and 132*). These trials were extended when further Morris 'Snubnose' vans were delivered in 1925 and 1926, together with Trojans

❖ 117

Above: This picture was taken c. May 1925 and shows a group of Morris and Morris-Commercial Royal Mail vans in use at Peterborough. The first three vehicles in the line, registered XX 9401, XX 9399 and XX 9398, are from the first production batch of six Morris 8 cwt chassis to be sold to the GPO. They were all erected on 26th March 1925 (see note 4, page 123). These three vans appear to be fitted with 105 cu ft bodies of similar design to the prototype – XR 1596 – as seen on page 132. Next in line are two Morris 8 cwt Standard Vans, which are probably from the Morris Motors Ltd. demonstration fleet, as they carry Oxford (BW) registration numbers. The last four vans in the line are part of the first batch of Morris-Commercial One-Ton 'T' types to be supplied to the GPO; the first 'Tonner' appears to be fitted with a 140 cu ft body whilst the bodies on the others are probably of 250 cu ft capacity

A 1927 Morris 'Snubnose' 105 cu ft Royal Mail van, standing alongside a de Havilland Rapide aircraft at Croydon Aerodrome in 1934. The vehicle, which had seen seven years of arduous service when photographed, was delivering letters for air transport to Liverpool and the North. It carried chassis no. 183801 and was one of a batch of 10 chassis (nos. 183800 to 183809) that were erected on 21st February 1927 with rear wheel brakes only to a plain specification. The entire batch was converted to four wheel brakes before they were despatched from the factory at Cowley

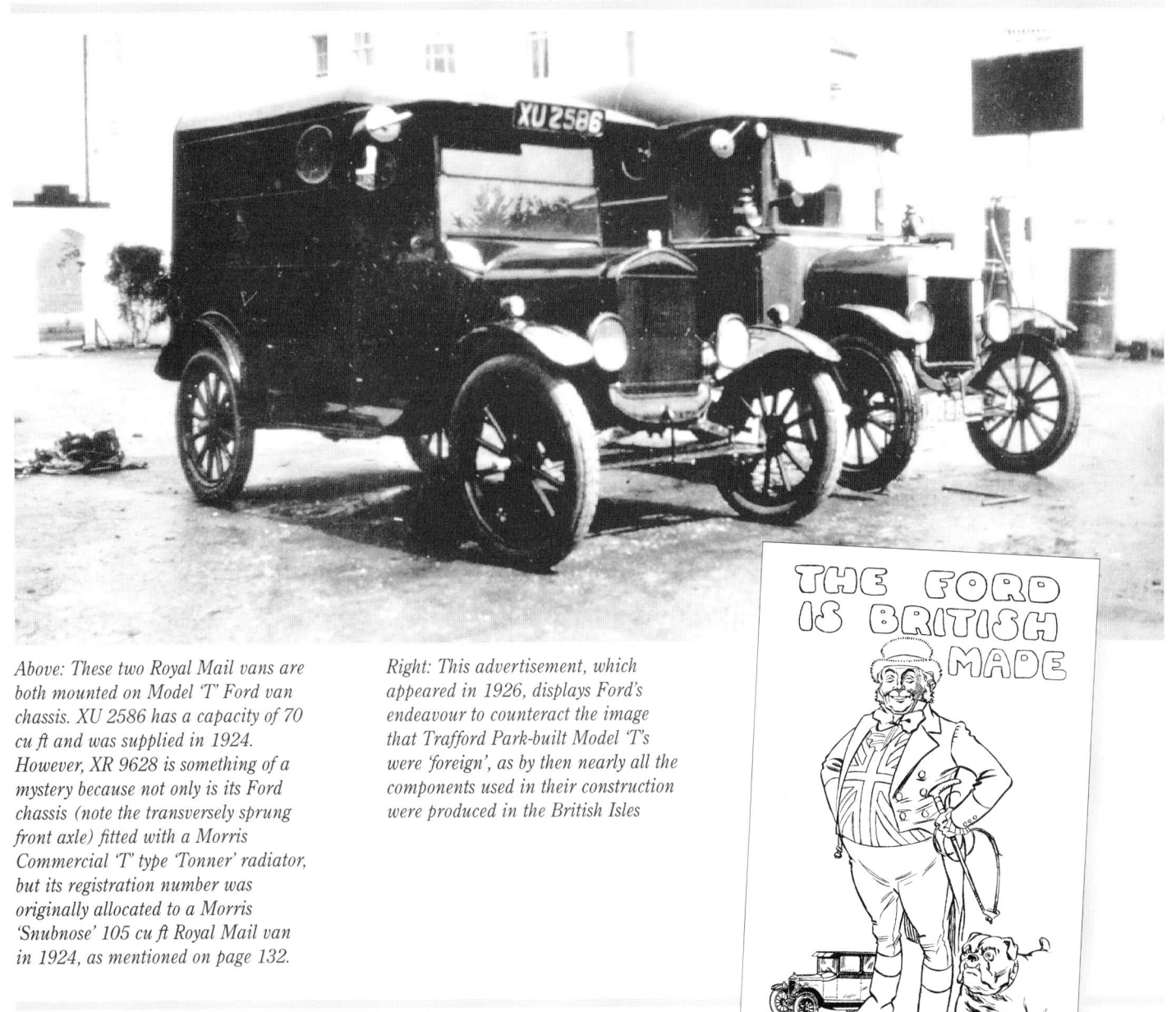

Above: These two Royal Mail vans are both mounted on Model 'T' Ford van chassis. XU 2586 has a capacity of 70 cu ft and was supplied in 1924. However, XR 9628 is something of a mystery because not only is its Ford chassis (note the transversely sprung front axle) fitted with a Morris Commercial 'T' type 'Tonner' radiator, but its registration number was originally allocated to a Morris 'Snubnose' 105 cu ft Royal Mail van in 1924, as mentioned on page 132.

Right: This advertisement, which appeared in 1926, displays Ford's endeavour to counteract the image that Trafford Park-built Model 'T's were 'foreign', as by then nearly all the components used in their construction were produced in the British Isles (*see page 128*) and Morris-Commercial 'T' type 'Tonners', but, at the end of 1927, the contest changed unexpectedly when the Model 'T' Ford was replaced by the more expensive Model 'A'. By the early 1930s, these events had culminated in Ford vans losing favour and Morris and Morris-Commercial chassis becoming the choice for all Royal Mail vans. The products of Morris Motors Ltd, and Morris Commercial Cars Ltd. were then to dominate the Royal Mail fleet to such an extent that by 1950 they represented 80% of the 23,000 vehicles within that fleet.

The chassis supplied to the GPO were usually equipped with bodies built to their own designs and specifications which were constructed by several different body builders. During the late 1920s and 1930s, the standard sizes of these bodies were 70, 105, 160, 250 and 340 cu ft capacity, with each increment having only subtle changes to create a fleet of similar appearance. These body sizes are significant because the GPO classified their mail vans by body capacity rather than by make and model name/payload.

The table on page 122 details the numbers of Ford, Morris and Morris-Commercial vans, as well as other makes, that were delivered annually to the GPO for their Royal Mail fleet from 1920 to 1935. The average life of a Royal Mail van was five to six years.

A 1927 season Morris 'Snubnose' 10 cwt High Top Light Van (see pages 71 & 72) pictured in the Despatch Department of the Morris factory at Cowley and finished in Royal Mail livery. This vehicle probably carries the chassis no. 160585, (a GPO 'special' erected on 5th October 1926) the first 10 cwt van to be made in the 1927 season and therefore the first van to be based on a 'Flatnose' Morris Cowley chassis and to have a large 'Snubnose' radiator (see page 46)

A 1934 Morris 'Flatnose' 70 cu ft Royal Mail van (serial no.5276) photographed in 1935 outside The Land's End Hotel in Cornwall. 'Flatnose' production officially ceased in 1933, but 164 Flatnose van chassis were built specially for the GPO during 1934, this chassis being one of them. Note the 5 stud wheels and 12" brake drums as specified by the GPO in lieu of the 3 stud wheels and 9" brake drums usually fitted to 'Flatnose' vans (see page 140) and that no front bumper is fitted even though the chassis has brackets to suit.

Deliveries of vans to the Royal Mail fleet – 1920 to 1935

Make of chassis/type & body capacity	1920/1	1921/2	1922/3	1923/4	1924/5	1925/6	1926/7	1927/8	1928/9	1929/30	1930/1	1931/2	1932/3	1933/4	1935
GWK Model 'F' 8 cwt, 80 cu ft	56[2]	100	1	–	–	–	–	–	–	–	–	–	–	–	–
Trojan[3]	–	–	–	–	11	1	–	12	–	–	–	–	–	–	–
Ford 70 & 105 cu ft[1]	4	58	87	60	116	208	173	37	2	–	75	–	–	–	–
Total Ford 70/105/140/ 250/320 cu ft	60[1]	200[1]	144	81	181	probably 278	225	53	4	–	90	–	–	–	–
Morris 8 & 10 cwt 70 cu ft	–	–	–	–	–	6	33	179	288	83	167	115	240	333[9]	50[10&11]
Morris 8 & 10 cwt 105 cu ft	–	–	–	2[4]	–	7[4]	43	145	–	–	–	–	–	–	–
Morris-Commercial 'L2/8', 70 cu ft	–	–	–	–	–	–	–	–	–	–	–	–	–	–	248[11]
Morris-Commercial 'L' & 'L2', 105 cu ft	–	–	–	–	–	–	–	1[5]	194	158	107[5]	127	142	308	216
Morris Minor 30/35 cu ft	–	–	–	–	–	–	–	–	–	–	–	12[6]	2	484	560
Total Morris & Morris Comm. 30/35/70/105/ 140/160/240/ 250/340 cu ft	–	–	–	2	1[12]	19	84	418	588	356	365	365	507[7]	1282[8]	1282

See Notes opposite

Notes on the table

1. Ford Model 'T' 7 cwt chassis were fitted with 70 & 105 cu ft bodies until 1927 when they were replaced by the Ford Models 'A' (24.03 h.p.) and 'AF' (14.9 h.p.) chassis, both of 10 cwt capacity (*see page 131*). Some, if not all, of the Fords supplied between 1921 and 1923 were reconditioned chassis ex-War Department.
2. The first GWK, purchased in February 1920 for trial, was second-hand (*see page 126*).
3. Trojans were manufactured by Leyland Motors Ltd. (*see page 128*).
4. The first Morris 8 cwt 'Snubnose' Royal Mail van was registered XR 1596 in 1924 (*see page 132*). It was soon followed by a similar vehicle, XR 9628. Following initial trials with XR 1596 and XR 9628, Morris Motors Ltd. supplied their first production batch of six 'Snubnose' 8 cwt van chassis to the GPO, which were all probably fitted with 105 cu ft bodies, as detailed below:

Chassis no.	*Erected*	*Despatched*	*Registration no. of van*
83703	26.3.25	31.3.25	XX 9399
83706	"	"	XX 9401
83707	"	"	XX 9398
83708	"	"	XX 9397
83709	"	"	XX 9400
83710	"	"	XX 9402

(Three of these Royal Mail vans are illustrated on p.118)

These chassis were fitted with 12 volt dynamotors (a combined starter and dynamo unit) but the next Morris 8 cwt van chassis to be delivered to the GPO (i.e. nos. 142191/2 and 142642 to 142648, which were all erected during April 1926) were fitted with the cheaper 6 volt dynamos and less Gabriel Snubber shock absorbers.

5. From 1928, 105 cu ft bodies were mounted on the Morris-Commercial 'L' type 12 cwt chassis. The first one was registered YU 2525. From 1931, 105 cu ft bodies were mounted on 15 cwt Morris-Commercial 'L2' chassis, which replaced the 'L'.
6. The first batch of 12 Morris Minor Royal Mail vans was introduced into the Royal Mail fleet during 1932 (*see page 124*). The Morris Minor van provided Morris Motors Ltd. at Cowley with continued volume business with the GPO after the manufacture of the 'Flatnose' Morris 8/10 cwt chassis had ceased.
7. Three Morris-Commercial 30 cwt vehicles with 340 cu ft bodies supplied during 1932/33 were second-hand.
8. In April 1933 the Royal Mail fleet consisted of 4,399 vehicles, of which 324 were Fords and 2,673 were Morris/Morris-Commercial. The balance was made up of Maudslays, Trojans, petrol/battery electrics and motor cycles/motor cycle combinations. The Post Office Telephones fleet also operated Morris vehicles (mostly 15 cwt, 1 ton and 30 cwt Morris-Commercials) and in April 1933 this fleet consisted of 3,359 vehicles, of which 26 were Fords and 495 were Morris/Morris-Commercials.
9. The last 'Flatnose' chassis, no 384370, was despatched from the factory on 22nd June 1934. The last batch of 32 'Flatnose' 70 cu ft Royal Mail vans was registered BGF 344 to BGF 375. A prototype Morris 1935 season (*see page 141, note 5*) 'TWV' van chassis, no. 384448, was registered BGP 660 in August 1934.
10. A batch of 50 Morris 8/10 cwt 'TWV' vans (similar to the prototype BGP 660, note 9) was made during November and December 1934. (*See page 141, note 5*).
11. From May 1935, Royal Mail 70 cu ft van bodies were mounted on Morris-Commercial 'L2/8' chassis thereby replacing the car derived Morris 8/10 cwt chassis. (*See page 143, note 6*). The first (prototype) Morris-Commercial 'L2/8' Royal Mail van, chassis no. 3228, was registered BUU 214.
12. The first Morris-Commercial Royal Mail van was delivered to the GPO in March 1925: a 'T' Type Tonner with a 240 cu ft body, registered XX 1257.

This picture was taken at the Cowley works of Morris Motors Ltd. during the early part of 1932 and shows the first 12 Morris Minor 5 cwt, 30 cu ft (later reclassified as 35 cu ft) Royal Mail vans to be manufactured for the GPO. They are shown awaiting inspection prior to delivery and were subsequently registered GW 1550 to GW 1553 and GW 6058 to GW 6065. This first batch is significant because the introduction of Morris Minors enabled the GPO to start using mailvans instead of motorcycles for the motorisation of collection and delivery of mails in rural areas. As a result, Morris Minor vans were soon to dominate the Royal Mail fleet, thereby generating substantial business for Morris Motors Ltd. who supplied some 3,660 of these vehicles between 1932 and 1940. In addition, some 3,719 Morris Minor vans were also supplied to the Post Office Telephones fleet over the same period

One of the batch of 12 Morris Minor Royal Mail vans shown above, pictured in 1932 after it had been allocated to the Bedford Head Postmaster and had commenced field trials

An advertisement for a 1932 season Morris Minor 5 cwt van

MORRIS LIGHT TRANSPORT

The MORRIS 5-CWT. VAN

Price (*ex Works*) .. £110

Triplex safety glass is fitted to windscreen and door-windows, and all external bright parts are chromium finished.

Built to meet the needs of tradesmen whose businesses require fast and mobile light transport of absolute dependability, the Morris 5-cwt. Van, with its highly developed economical 8 h.p. side-valve engine, generous chassis specification, and extraordinarily good performance, provides both small and large tradesmen with business-expanding transport. The spacious body possesses an interior height of 3 ft. 6 in., a width of 3 ft. 4½ in., and a length behind the driver's seat of 3 ft. 4 in., giving the useful capacity of 38 cubic feet. The equipment is very complete, and the finish is in shop grey in readiness for the finishing colour of your choice.

GWK Royal Mail Vans

Background and difficulties

The initials GWK arose from the founders of the business, Arthur Grice, J. Talfourd Wood and C. M. Keiller who had planned a new light car during 1910. The following year a company was formed and operations were transferred from their workshop in Beckenham to Datchet, near Windsor, where their Models 'A' and 'B' were produced. These cars were fitted with a rear mounted Coventry-Simplex vertical twin cylinder engine, a friction drive transmission and had a chassis weight of $6^1/_2$ cwt, which bridged the gap between true cyclecars, like the $4^1/_2$ cwt GN, and the $8^1/_2$ cwt 'White & Poppe' Morris Oxford.

The principles of the friction transmission were simple. Two discs, one connected to the engine and the other to the road wheels, were mounted at right angles to each other. The driven disc, which was shod with a 'tyre', could slide across the driving disc and engage in four different positions, so that the ratio between the engine and the road wheels could be varied. De-clutching moved the driven disc out of contact with the driving disc.

Although GWK tried several different types of material for the 'tyre' of the driven disc, notably millboard, Ferodo and cork, slippage was often evident but could usually be overcome by keeping the discs free from oil or grease. However, the main problem with the transmission occurred when the clutch was let in too slowly or when driving with the handbrake on, which resulted in flats being worn on the 'tyre'. GWK recommended that the clutch should be let in more quickly than a cone or plate type but this was not always possible when starting on gradients, especially with a fully-loaded vehicle. Although GWKs could be driven with flats on the 'tyre' of the driven disc, a severe vibration and knocking could be set up.

Despite the problems with the friction transmission, about 1000 GWK models 'A' and 'B' were made before the First World War. After the war a new company, GWK (1919) Ltd., was set up in Maidenhead. The Model 'F' was then introduced which had a front-mounted 4 cylinder side valve 1368cc Coventry-Simplex engine, friction transmission and quarter elliptic suspension. At $10^1/_2$ cwt, and with a wheelbase of 9' 6", the chassis was bigger and heavier than the earlier models.

The Model 'F' proved to be problematic and Wood and Keiller worked hard to rectify such things as wheel-shedding, transmission whip and noisy transmissions. Whilst this was going on, Arthur Grice obtained a contract from the GPO for the supply of 50 vans, based on the Model 'F' with a capacity of 80 cu ft and a payload of 8 cwt, following trials with a second-hand vehicle. These vans were delivered between June and October 1920 and a further 106 were supplied during 1921/22. Not only were the vans found to be unreliable, but the Post Office drivers were unable to cope with the idiosyncrasies of the friction drive and by 1925 all 157 GWKs had been withdrawn from the Royal Mail fleet.

The protracted moulder's strike exacerbated GWK's problems and in 1922, GWK (1919) Ltd. went into liquidation and no further GWK vehicles were made.

Between 1920 and 1922, the GPO purchased 157 GWK Model 'F' Royal Mail vans and the three shown here were amongst the first 50 to be supplied. By 1925 all the GWK vehicles had been withdrawn as they were found to be unreliable

Below: The transmission of a Model 'F' GWK showing its friction drive arrangement

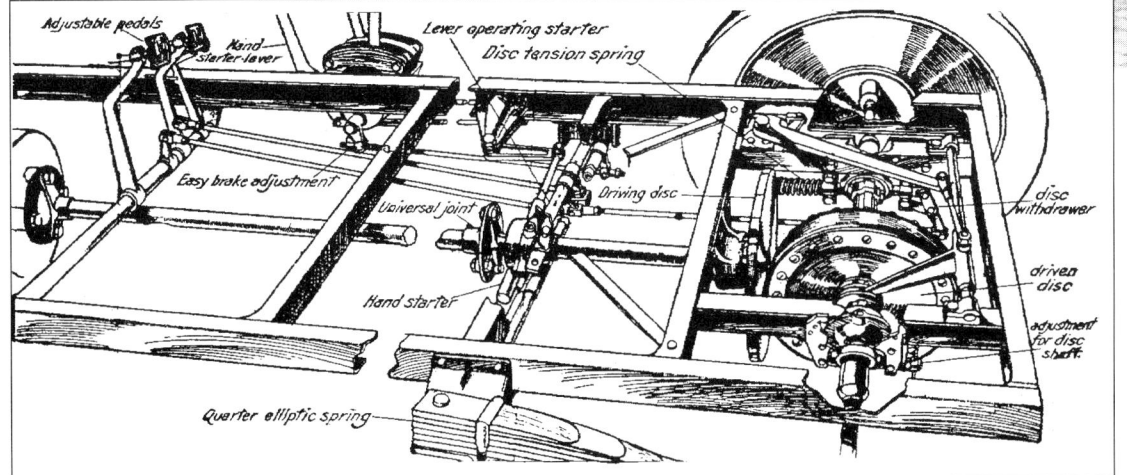

Trojan Royal Mail Vans

The 24 Trojans which were introduced into the Royal Mail fleet between 1925 and 1927 for trial purposes, were made, on a royalty basis, by Leyland Motors Ltd. at their Ham Works, Kingston-upon-Thames, Surrey.

Athough the vehicles were reliable and simple to maintain, their design was unconventional. The 1527cc engine, which was started manually by the driver whilst seated, was a horizontal two-stroke 'square four' with each pair of cylinders sharing a common combustion chamber and a single sparking plug.

The engine, which was located under the seat, was mounted on a punt-type chassis frame; the area under the bonnet merely housing the petrol tank, carburetter and part of the steering mechanism. Drive to the solid (i.e. without differential) rear axle was via a 2-speed epicyclic gearbox and duplex roller chain. With a maximum power output from the engine of only 11 b.h.p., the Trojan was limited to a top speed of about 30 m.p.h. Nevertheless, the engine had a low petrol consumption and such good low speed torque that the van could climb the steepest hills, albeit very slowly.

The Trojan's braking system was designed in such a way that the handbrake lever operated a brake on the transmission whilst the footbrake pedal actuated a single drum brake on the rear axle.

Finally, there was the controversial feature of the cheap but durable solid tyres. Although the long cantilever springs enabled the Trojan to ride quite smoothly over poor road surfaces, the narrow section of the tyres sometimes caused the driver to get into difficulties when the wheels of the vehicle were caught in tramlines. This gave rise to a wag's quip that Trojans then had to go willy-nilly to the tram depot! In fact, it was usually possible to get the front wheels free of the tramlines by giving the steering wheel a sharp 'snatch'.

The GPO probably rejected Trojans for their Royal Mail fleet because they were slow and too unorthodox. Nevertheless, by 1933 the Telephone Engineers had 108 Trojans in their fleet – mostly 15 cwt utilities.

A Trojan chassis fitted with a 70 cu ft body for the Royal Mail. Note the solid, narrow section tyres

Trojan Service

Value for your money is emphasised in every feature of the Trojan Van. A Trojan Van will give long and faithful service costing you little to purchase and maintain, and adding every day to the increased business resulting from efficient delivery.

LOW PETROL CONSUMPTION. The petrol consumption up to 35 m.p.g. is remarkably low for a light commercial vehicle, and in view of the Petrol Tax is an important consideration. Also the simplicity of the engine keeps down the cost of running repairs and eliminates the need for a skilled driver.

PRICES
From - - £135

Write to-day for the name and address of your nearest agent, also for Illustrated Folder T.F.5 giving full range of vehicles and deferred terms on which they can be purchased.

Trojan

Purley Way, Croydon
(Sole Concessionaires in Great Britain for the Sale and Service of Trojan vehicles).
Manufactured by
LEYLAND MOTORS, LTD.

(Flynn's)

Eleven Trojan Royal Mail vans, with 70 cu ft bodies, were supplied during 1925 and this picture shows one of them (XX 4471) when it was a few months old. The vehicle is parked in Livingstone Road, Thornton Heath, Surrey, with its Postman/driver, Mr Reginald (Jack) Wood alongside, who had joined the GPO as a Telegram Boy in 1910. Shortly before this picture was taken, Mr Wood had been instructed to collect a van from the Leyland works at Kingston-upon-Thames, despite not having had any previous driving experience. On arrival, the works foreman gave Mr Wood a half-an-hour driving lesson before he took charge of the new van – at that time a driving licence could be obtained without the need for a test. Nonetheless, Mr Wood, whose son and granddaughter also joined the GPO, continued to drive mailvans for the next 31 years, without accident. This period included the hazards of the World War Two 'black-outs' and post-war 'smogs'

Ford Model 'T' Royal Mail Vans

The Model 'T' Ford, which was introduced at the end of 1908, was the first vehicle of any kind to be mass produced on a moving production line. Such production methods enabled very low prices to be achieved and these prices were the prime reason for the Model 'T's enormous popularity worldwide, resulting in over 15,000,000 examples being made by 1927, when it was replaced by the Model 'A'. Part of the success of the Model 'T' can also be attributed to the high grade materials used in its construction, which made it robust and durable.

Although its engine was of orthodox side valve design, having 4 cylinders with a capacity of 2896cc (22.4 RAC h.p.) and developing its maximum power at 1800 rpm, the Model 'T' had the anachronistic feature of a 2-speed epicyclic gearbox, which could be handled more easily by unskilled drivers than the conventional 'crash' gearbox. Top gear gear ratio was 3.46:1 which produced, with the remarkable flexibility of the engine, a speed range of 3 to 45 m.p.h., while bottom gear, with a ratio of 10:1, gave a maximum speed of 15 m.p.h. Like the rest of the vehicle, the gearbox proved to be reliable in service but when its oil was cold, so much 'drag' was created that it was almost impossible to start the engine by hand. Consequently, before the starting handle was turned on a cold morning, one rear wheel needed to be jacked up, and the handbrake released, to allow the transmission to revolve freely.

With three foot pedals, the controls of the Model 'T' Ford appeared normal, but their operation was as follows: the right-hand pedal activated a transmission brake, the central pedal operated reverse gear and the left-hand pedal selected the two forward speeds and neutral. The throttle was controlled by a hand lever mounted under the steering wheel, while a similar lever controlled the ignition advance/retard. The initial application of the hand brake lever automatically moved the left-hand foot pedal to its neutral position, while further application of the lever operated the rear wheel brakes.

In 1911 the Ford Motor Company opened a factory at Trafford Park in Manchester to assemble Model 'T's with parts imported from the USA. Ford had chosen Manchester because its port offered more

Above: The controls of a Model 'T' Ford

Below & top right: Model 'T' Ford 7 cwt chassis fitted with 105 cu ft bodies for the Royal Mail

Right: By extending the chassis and fitting a stronger chain driven rear axle, this Baico conversion of a Model 'T' Ford chassis raised its payload to 30 cwt. BAICO is a contraction of 'British American Import Co.'

competitive freight rates and services than its rivals. The company remained at this site until 1931 when its operations were transferred to Dagenham.

Ford expanded production at Trafford Park during the First World War whilst British manufacturers were engaged on armament contracts and other war work. When peace returned in 1918, Ford could immediately take advantage of the post war boom because, unlike most other manufacturers, their Trafford Park factory did not have to be reorganised to make vehicles.

Consequently, during 1919 and 1920, Ford dominated the British market by selling 58,355 Model 'T's, of which 9,989 were 7 cwt Delivery Vans.

Between 1921 and 1926 the GPO purchased over 1,000 Ford Model 'T' (7 cwt) and 'TT' (1 ton) chassis and those which operated within their Royal Mail fleet were fitted with 70, 105, 140, 160, 250 and 320 cu ft bodies. The 320 cu ft body was mounted onto a chassis that had been lengthened and fitted with a stronger rear axle, driven by the original transmission via chains and sprockets that were arranged to give a lower overall gear ratio. These modifications were carried out by Baico Patents Ltd. of Fulham Road, London.

After the demise of the Models 'T' and 'TT' in 1927, some Ford Models 'A'/'AF' (10 cwt) and 'AA' (1 ton) chassis were supplied to the Royal Mail to suit their range of bodies until 1931, when Ford products lost favour to those of Morris Motors Ltd. and Morris Commercial Cars Ltd.

Morris & Morris-Commercial Royal Mail Vans

The Morris 105 cu ft Royal Mail Van, 'XR 1596'

XR 1596 was the first of two* 105 cu ft vans to be supplied by Morris Motors Ltd. to the GPO in 1924 for trial and this van deserves special mention because it paved the way for the supply of thousands of Morris and Morris-Commercial vehicles to the GPO for their Royal Mail and Post Office Telephones fleets. The chassis of XR 1596, which was similar to the contemporary 'Bullnose' Morris Cowley car except for its rear road springs and 'Snubnose' radiator, carried the number 33939. It was erected at Cowley on 12th October 1923† with an 11.9 h.p. type CB engine (no. 42178), and was road tested on the same day. The chassis was built to a 'plain' specification, i.e. with a 6 volt dynamo in lieu of a dynamotor and in common with all Morris vehicles made at this time, the wheels were shod with beaded edge tyres.

Shortly after being erected, the chassis was sent to a bodybuilder for the construction of a 105 cu ft body to the GPO's specification and it then returned to Cowley where the complete vehicle was wired on 19th November 1923. Subsequently, the van received two 'final' tests, the first on the 2nd December 1923 and the second on 22nd February 1924. Presumably Morris Motors Ltd. wanted to make absolutely sure all was in order before it was despatched from the factory on 28th March 1924. 'XR 1596' then commenced service and trials with the Post Office at Bangor during April 1924.

---

* The second Morris 105 cu ft Royal Mail van to be supplied was registered 'XR 9628'.

† Although Morris Motors Ltd. announced their 'Snubnose' vans at the end of November 1923, they had been making them since June 1923, in order to build up stocks prior to announcement.

Left and above: This Morris 8 cwt 'Snubnose' Light Van chassis, fitted with a 105 cu ft body and registered XR 1596, was the first Morris Royal Mail van to be supplied to the GPO. The chassis was erected on 12th October 1923 (chassis no. 33939) and the vehicle was handed over to the GPO on 28th March 1924, to commence service and trials with the Post Office at Bangor. It paved the way for the supply of many thousands of Morris and Morris-Commercial vehicles to the GPO

Below: A 1929 Morris Commercial 'L' type 105 cu ft Royal Mail van having a three lamp (electrical) set and two oil lamps

Above: 1929 'Snubnose' Morris 10 cwt Light Van chassis fitted with a 70 cu ft body for the Royal Mail. Unlike the 1927 season Royal Mail van shown on page 117, this vehicle has a 5 lamp (electrical) set, whilst retaining the two oil lamps, and has side screens installed. Note the spare wheel mounting bracket, located between the offside door and rear wheel arch (see page 138, note 1)

This Morris Commercial 'L' type 105 cu ft Royal Mail van, which was registered in 1930, is seen being unloaded at the Queens Ferry jetty adjacent to the Forth Railway Bridge in Scotland. The photograph was taken on 16th August 1934

POST OFFICE
D. MACGREGOR & SONS
MERCHANTS
TELEPHONE
ROYAL MAIL
GR
CADBURYS CHOCOLATE
KENSITAS
KENSITAS
POST OFFICE

This 1932 'Flatnose' Morris 70 cu ft Royal Mail van was photographed (left) in 1934 in the village of Drymen on the borders of Dumbarton and Stirling and (right) on the shores of Loch Lomond. The vehicle remained in service until January 1939. Note the 5 stud wheels and 12" brake drums as specified by the GPO in lieu of the 3 stud wheels and 9" brake drums, usually fitted to 'Flatnose' vans

Morris & Morris-Commercial 70 & 105 cu ft Royal Mail Van chassis supplied between 1924 to 1935

Model	Wheelbase	GPO body in cu ft	Date	Note no.
Morris 'Snubnose' 8 cwt (small radiator)*	8' 6"	70 & 105	1924 & 1926	1
Morris 'Snubnose' 10 cwt (large radiator)*	8' 9"	70 & 105 70 only	1927–28 1928–30	1 & 3
Morris-Commercial 'L' type, 12 cwt†	9' 6"	105	1928–30	2
Morris 'Flatnose' 8 cwt*	8' 9"	70	1931–34	3 & 4
Morris Cowley 'TWV', 10 cwt*	8' 9"	70	1935	5
Morris-Commercial 'L2', 15 cwt†	9' 6"	105	1930–38	6
Morris-Commercial 'L2/8', 15 cwt†	8' 6"	70	1935–38	6

* 'Snubnose', 'Flatnose' and 'TWV', 8 & 10 cwt van chassis were manufactured by Morris Motors Ltd.
† Morris Commercial 'L'. 'L2' & 'L2/8' van chassis were manufactured by Morris Commercial Cars Ltd

Notes

1 The 8 cwt chassis supplied between 1924 and 1926 were derived from the 'Bullnose' Morris Cowley, while the 10 cwt chassis supplied from 1927 were derived from the 'Flatnose' Morris Cowley. Both types were fitted with either 70 or 105 cu ft bodies but from 1928, the 105 cu ft body was mounted onto the 12 cwt Morris-Commercial 'L' type, and later the 15 cwt 'L2' chassis, leaving the 'Snubnose' 10 cwt chassis to be fitted with the 70 cu ft design only. Perhaps the quantity of letters and parcels that could be carried in a 105 cu ft body overloaded the 'Snubnose' 8/10 cwt (payload) chassis and axles.

Originally, the spare wheel for 'Snubnose' Royal Mail vans had been carried in the van interior, nearside, adjacent to the sliding panel. In this position, it was found that the wheel (i) obstructed entrance

A 1931 Morris-Commercial L2 105 cu ft Royal Mail van having the side lights mounted above the doors and oil lamps

to the body of the van by way of the sliding panel, (ii) reduced the carrying capacity of the van and (iii) in the event of a puncture, the spare wheel could not be easily removed when the van was fully laden, so unless the driver received assistance, he had to run on a flat tyre until a secure place could be found to unload the van.

A decision was then taken in February 1929 to mount the spare wheel onto the offside of the scuttle below the windscreen upright. Although the spare wheel restricted the opening of the driver's door in this position, this was not considered to be a problem and in fact the GPO preferred drivers to alight from the nearside door of their vans as this was thought to be safer. Nevertheless, two other positions had been considered (i) under the rear of chassis; but there was not sufficient overhang or clearance, and (ii) on the roof; but its construction was not designed to carry any weight.

However, following an accident, where the van overturned and came to rest on its nearside, the driver became trapped within his vehicle as he was unable to open the offside door. It was then decided to mount the spare wheel between the offside door and the rear wheel arch, (*see pages 134 & 151*) thus removing any restriction to the opening of the driver's door.

It seems that the left-hand front wheel of Morris 'Snubnose' Royal Mail vans vans were often fitted with a hub cap housing an odometer. Several types of hub cap odometers were available from accessories manufacturers, an example of which is shown here.

2 The design of the Morris-Commercial 'L' type 12 cwt chassis, differs entirely from that of the Morris 'Snubnose' and 'Flatnose' chassis, but for the 11.9 h.p. type CB engine and gearbox. Compared to a 'Snubnose'/'Flatnose', the 'L' type has a more robust chassis frame and axles (the rear is worm drive) and a radiator with a cast aluminium surround.

3 Although the 8 cwt 'Flatnose' Morris van chassis introduced in 1931 were similar to the earlier 'Snubnose' 10 cwt van chassis, except for the radia-

❖ 139

tor, they were fitted with a more powerful 13.9 h.p. type CG engine (from engine no. 341552, Sept. 1930), which incorporated a 'turbulent' (Ricardo) type of cylinder head. However, GPO 'Flatnose' van chassis could be fitted with the 11.9 h.p. type CB engine as an option, although this type was replaced c. November 1931 by an 11.9 h.p. type CK with cast iron pistons, as detailed later.

Certain changes to the standard specification were made to the 'Snubnose' and 'Flatnose' van chassis supplied to the GPO, and those changes that are known are summarised as follows:

Pistons

Although aluminium pistons had been fitted as standard since 1925, the GPO called for 'lightweight' cast iron pistons (pt no. CK2122, 11.9 h.p. & CG2550, 13.9 h.p.) to be installed in engines fitted to their Morris Royal Mail vans. These pistons, which were introduced at engine no. 404145 (c. November 1931) had three rings, of which one was a scraper. At the request of the GPO's Engineer-in-Chief, Morris Motors Ltd. wrote to their dealers in May 1932 to ensure that engines installed in Royal Mail vans were rebored and fitted with pistons in accordance with GPO requirements. Dealers were also asked to identify engines which belonged to the GPO when they were returned to the factory for reconditioning.

Front and rear axles

From c.1931, 'Flatnose' Morris Oxford (car) front and rear axle assemblies, with 5 stud wheels/hubs and 12" brake drums were fitted to 'Flatnose' Morris van chassis supplied to the GPO. These axles were installed in lieu of 3 stud wheels/hubs and 9" brake drums which were the standard fitting to both 'Flatnose' Morris Cowley (cars) and light vans. In addition to having larger brakes, axles with 5 stud wheels/hubs had a more robust rear hub assembly and wheel bearings with a double row of balls (as opposed to a single row on 3 stud rear axles), which probably enabled the payload of 'Flatnose' 70 cu ft Royal Mail vans to be classified as 10 cwt, instead of 8 cwt as standard.

Rear axle half shafts

Special half shafts were called for by the GPO. These shafts were made of different steel and they also received different heat treatment from standard. They were identified by dabs of yellow paint.

Road springs

Morris Standard van 1928 to 1931

	Pt no.	Free camber	No. of leaves/Total thickness
Front	2694	2.375"	5 + 2 rebound/1.4777"
Rear	2155	4.25"	8/2.125"

Morris Royal Mail van 1928 to 1931

	Pt no.	Free camber	No. of leaves/Total thickness
Front	1922[1]	2.00"	6 + 2 rebound/1.7085"
Rear	3272	4.25"	8/2.125"

1 Common to 1926/7, 'Flatnose' Morris Cowley/Oxford cars.

A service letter sent to Morris Dealers in 1930 reminds them to reset the road springs fitted to Royal Mail vans to GPO specifications.

Right: A 1935 Morris-Commercial L2 105 cu ft Royal Mail van, serial no. 10,000 (later renumbered 1655), with trafficators and side lamps mounted on the door pillars

Right: A 1937 Morris-Commercial L2, 105 cu ft Royal Mail van, serial no. 12048. Compared with 'BXC 285' above the oil lamp brackets have been lowered to enable the side lamps to be repositioned

Brake linings

Wire bonded brake linings were fitted to Royal Mail vans. These linings had to be cut to length from rolls and then drilled/countersunk to suit the brake shoes.

Radiator drain taps

Because the GPO did not use 'anti-freeze', the cooling system of their vehicles was filled and drained daily. Consequently, heavy duty drain taps were called for.

Dynamos

Royal Mail vans were usually fitted with dynamos, initially the 6 volt Lucas E35 and later the Lucas 12 volt E418, in lieu of Lucas dynamotors (a combined starter and dynamo unit).

4 After 'Flatnose' Morris Light Vans were withdrawn from general sale at the end of the 1933 season (at chassis no. 382397, which was erected on 19th July 1933) 164 'Flatnose' van chassis were manufactured during the 1934 season specially for the GPO. The last GPO 'Flatnose' van chassis and therefore the last 'Flatnose' to be built, was chassis no. 384370, which was erected on 5th June 1934. This chassis was despatched from the factory on 22nd June 1934.

The last batch of 32 'Flatnose' Morris 70 cu ft vans to be supplied to the GPO was registered BGF 344 to BGF 375 inclusive, during the summer of 1934. They carried the serial nos. 5347 to 5378.

5 Following the withdrawal of the 'Flatnose' Morris 10 cwt Royal Mail van chassis, Morris Motors Ltd.

Left & opposite: Both sides of a 1934 Morris-Commercial L2, 105 cu ft Royal Mail van, serial no. 5216, photographed in the yard at King Edward Building, London. The chassis of the L2 and L2/8 (shown below) are similar except for their wheelbase (see note 6, page 143)

Right: A 1938 Morris Commercial L2/8, 70 cu ft Royal Mail van, serial no. 13933. This vehicle was one of the last L2/8s to be supplied before the introduction of the L3/8 early in 1939. Note the white edges of the wings and running boards to comply with wartime blackout regulations, and the distance between the door and rear wheel arch when compared with the 105 cu ft body mounted on an L2 (above and right). The L2/8 and L3/8 were specially made by Morris Commercial Cars Ltd. to suit the GPO's 70 cu ft body and replaced the Morris 'Flatnose' van chassis (see page 143, note 6)

then supplied a single 1935 season Morris 'TWV' van chassis to the GPO for their 70 cu ft body. (A 1935 season 'TWV' van is illustrated on page 167). This chassis, no. 384448, which was erected on 22nd June 1934, was probably a 'TWV' prototype. It was fitted with a type 'TF' engine (no. 3629) and registered BGP 660 during August 1934.

No further van chassis to suit the GPO's 70 cu ft body were manufactured by Morris Motors Ltd. until November and December 1934 when a batch of 50, 1935 season, Morris Cowley 'TWV' van chassis was erected for the GPO. They were issued with a set of 6 digit chassis numbers, i.e. 384518 to 384567,* that did not have a prefix to indicate year of manufacture and model, (e.g. 35/TWV 501), which was the system that had been adopted by Morris Motors Ltd. for vans since the beginning of the 1935 season.

It is thought that these 50 'TWV' vans were issued with a special set of chassis numbers because they were non-standard; their engines were probably 13.9 h.p. type CS, or 11.9 h.p. type CK, with 3 speed gearboxes, (as opposed to engine type TF with a 4 speed gearbox fitted as standard to 1935 season 'TWV' vans) to enable the GPO to have common power units within much of their fleet.

No further 'TWV' Morris 10 cwt van chassis were supplied to the GPO and the GPO then favoured the Morris-Commercial 15 cwt 'L2/8' chassis for their 70 cu ft body.

6 The 9' 6" wheelbase Morris-Commercial 15 cwt 'L2' was introduced in 1930. It was developed from and replaced the Morris-Commercial 12 cwt 'L' type with which it shared many similar components. However, the 'L2' was fitted with a larger type CP or CQ, 13.9

---

* These vans were registered BLH 572 to 581, BLH 583 to 600, BLR 101 to 106 and BLR 193 to 208.

h.p. engine, and a radiator with a chromium plated surround.

The 15 cwt 'L2/8', which was introduced in May 1935, was specially made for the GPO and was, in effect, an 'L2' shortened to a wheelbase of 8' 6" to suit the GPO's 70 cu ft body. The 'L2/8' replaced the Morris 'Flatnose' and 'TWV' van chassis within the Royal Mail from 1935.

Because the 'Flatnose' and 'L2/8' Royal Mail vans both have 5 stud wheels and are fitted with similar bodies, it is sometimes difficult to distinguish them in side view photographs when the radiator cannot be seen. However, there are two bonnet catches fitted to the top of the bonnet side panels on 'L2/8's to enable the side panels to be detached from the top panel. Because 'Flatnoses' have a one-piece bonnet they do not have bonnet catches at the top of the side panels, thereby providing a useful distinguishing feature.

The first (prototype) Morris-Commercial 'L2/8' to be supplied to the GPO carried the chassis no. 3228 – it was registered BUU 214 during February 1935. The first 'production' 'L2/8's were delivered during May 1935, when a batch of 22 vans was registered BXC 291 to 312 – they were built under sanction no, 521 with the chassis nos. 3521 to 3525, 3531 to 3535 and 3538 to 3549 inclusive.

There are certain mechanical differences between the 'standard' 'L2' and the GPO 'L2' and 'L2/8', which are summarised as follows:

	GPO L2 & L2/8	'Standard' L2
Carburetter	Solex*	Smiths Single Jet type 26HK
Ignition	Automatic advance device incorporated in distributor	Manual advance/retard lever, mounted on steering column
Foot brakes	Hydraulic (L2/8 only)	Rod operated
Electrical	Wiring modified to include an inspection lamp plug	

The hydraulically-operated brakes on the L2/8 called for redesigned front hubs and brake drums which resulted in an increased front track. However, the front axle and stub axles were common to both the L2 and L2/8.

---

* Two types of Solex carburetter were fitted – one type includes a thermostarter device and a governor whilst the other type includes a thermostarter device but with no governor. The thermostarter provides an automatic choke and it is in effect an auxiliary carburetter that is brought into action when the engine is cold, by means of a bi-metallic strip, to give a rich mixture. The governor arrangement calls for an additional butterfly, which is offset with a spring to hold it open. When the air speed across it reaches a certain velocity, the offset overcomes the spring and the butterfly closes, hence restricting both the maximum speed of the vehicle and the maximum revs. of the engine. By providing an automatic choke and an automatic ignition advance, manual controls for these items are eliminated on GPO 'L2' and 'L2/8's, thereby preventing driver misuse and at the same time ensuring that the engine is operating as economically as possible.

Above: One of the initial five 1933 Morris-Commercial 15 cwt L2s, registered 'JJ 1143', with bodies manufactured by Carter, Paterson & Co., Ltd (Carriers), for the Post Office Telephones fleet

The Morris-Commercial L2 and L2/8 were produced in large numbers to meet the demands of both the Royal Mail and the Post Office Telephones fleets. They were replaced in 1938 by the Morris-Commercial L3 and L3/8 (*see pages 110 & 113*).

This advertisement appeared in October 1934

The G.P.O. stage the biggest telephone DRIVE on

in their history. This thoroughly EFFICIENT Government Department—amongst the keenest buyers of transport in the country—rely for 100% EFFICIENCY in their delivery services—

MORRIS-COMMERCIAL
BRITISH TO THE BACKBONE

The G.P.O. already employ over 2,500 Morris-Commercial Vans

Ensure the utmost efficiency in your Transport — write for full particulars of the Morris-Commercial range of vehicles—15 cwt. to 70 cwt.

MORRIS COMMERCIAL CARS LTD.
ADDERLEY PARK, BIRMINGHAM, 8

YOU'RE SAFE WITH MORRIS-COMMERCIAL

Specifications for Morris-Commercial 'L' 12 cwt chassis, 'L2/8' & 'L2' 15 cwt chassis

	Engine	Carburetter	Ignition	Clutch	Gearbox	Steering	Prop. shaft	Rear axle ratio	Brakes	Turning circle	Wheelbase	Track	Wheels	Tyres	Nett chassis weight (approx.)
'L'	type CB 11.9 h.p. (1548cc)	Smiths single jet type 26 HKMC	Lucas magneto	Twin plate cork insert	3 speed & reverse	Worm & wheel	Enclosed by a torque tube. Late models – open	7.33:1 std. 6.25:1 option	14" dia. rod operated	45' rh 42' lh	9' 6"	4' 8" F&R	Artillery well-base	31×4* F=60psi R=50psi	–
'L2/8'	type CP or CQ 13.9 h.p. (1802cc)	Solex with auto choke (see text)	Coil & distributor (see text)	"	4 speed & reverse	"	Enclosed by a torque tube	5.50:1	Hydraulic	38' 8"	8' 6"	Front 4' 8⅝" Rear 4' 8" (see text)	Artillery well-base	5.00×20 F=28psi R=36psi	17¾ cwt
'L2'	"	Smiths single jet type 26 HK std. Solex with auto choke GPO (see text)	Coil & distributor (see text)	"	3 speed & reverse†	"	"	6.25:1 std. 5.50:1 option	Rod operated	42'	9' 6"	4' 8" F&R	Well-base wire wheels std. Well-base artillery option	3.25×20 for wire wheels. 30×5.50* for artillery wheels GPO 5.50×20 F=26psi R=38psi	19 cwt

* overall diameter of tyre and section width
† 4 speed & reverse from Dec. 1932

A Morris Commercial L2/8, 70 cu ft Royal Mail van BXK 627, new in April 1935, and a 1929 season 'Flatnose' Morris Cowley saloon, parked outside a main gate of the Austin Motor Company's factory in Bristol Road, Birmingham.

Hotchkiss/Morris 11.9/13.9 engine types used in 70 & 105 cu ft Morris Royal Mail Vans – 1924 to 1940

The Hotchkiss/Morris 11.9 and 13.9 h.p. engines were derived from the American Continental Red seal type 'U' engine, which is illustrated on page 35

Date	Model	Engine type
1924–30	'Snubnose', 8/10 cwt, 70/105 cu ft	CB, 11.9 h.p.
1931–34	Morris 'Flatnose', 10 cwt, 70 cu ft	CK, 11.9 h.p. or CG, 13.9 h.p.
1935	Morris 8/10 cwt 'TWV' 70 cu ft	(See note 5, page 141)
1929–30	Morris-Commercial L, 105 cu ft	CB 11.9 h.p.
1931–39	Morris-Commercial L2.8/L2, 70/105 cu ft	CP & CQ 13.9 h.p.
1939–40	Morris-Commercial L3.8/L3, 70/105 cu ft	CSDC, 13.9 h.p.

All these engine types share the same serial no. sequence.

(*See page 40 for engine specifications*)

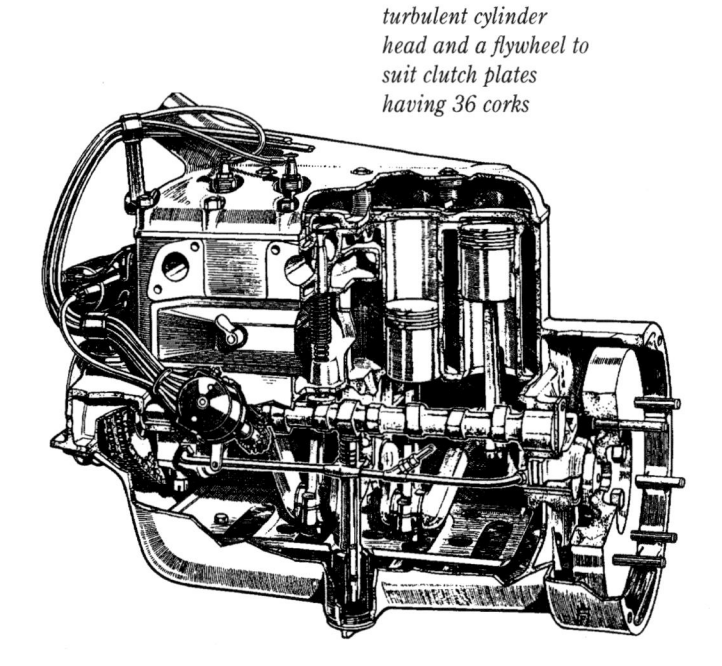

Engine Type CQ, 13.9 h.p., having a chain driven camshaft, coil ignition with a distributor driven by the camshaft, non-turbulent cylinder head and a flywheel to suit clutch plates having 36 corks

Right: Two 70 cu ft Royal Mail vans – a 1928 Morris 'Snubnose' (YU 8413) and a 1933 Morris 'Flatnose' – standing outside Haslemere Sorting Office. Note the three lamp (electric) set on the 'Snubnose' and five lamp (electric) set on the 'Flatnose' with side lamps mounted at the top of the windscreen. (photo: Mrs Cook)

Left: 1933 Morris 'Flatnose' Royal Mail vans with 70 cu ft bodies pose outside Weston-Super-Mare Post Office in the summer of 1933. These were the first Morris vans to be used at this post office, the previous vehicles having been Model 'T' Fords. (photo: Mr. Priddle)

CUSTOMS 5
IMPERIAL AIRWAYS
EXPORT
AIR MAIL
AIR MAIL
GJ 5159
ROYAL MAIL
GR

Three views of 'GJ 5159', a Morris 10 cwt chassis (no. 336680) with a 70 cu ft body, the first blue Royal Mail van to be used in the United Kingdom. The photograph on the left was taken c.1930 outside the Customs Hall, and the one below, in front of the main terminal block of Croydon Aerodrome which had been rebuilt in 1928 to cater for the ever-growing volume of air traffic. 'GJ 5159', which was new in May 1930, operated between London and Croydon Aerodrome carrying air mail. Note that the top right photograph shows a front wheel hubcap which houses an odometer (see page 139)

12

Relationships Between the GPO, Morris Motors Ltd. and their Dealers

W. R. Morris must have been delighted when the GPO purchased more Morris vehicles than Ford, for the first time, during 1927/28. To maintain this momentum Morris dealers were asked in a circular letter to give a 'Priority of service' to Royal Mail vans. However, a few dealers apparently saw the GPO as a source of easy revenue as the factory received complaints that some were recommending expensive engine overhauls at low mileages. This prompted Morris Motors Ltd. to send another circular letter at the beginning of 1929 which asked dealers to 'keep running costs as low as possible [on Royal Mail vans] in order that future business may develop'.

Subsequently, Morris's Dealer Contracts spelt out the right of the factory to deal direct with the GPO both in supplying vehicles and parts, and in particular the very high discounts (irrespective of quantity) that dealers had to give to the local GPO workshops.

Right: A Morris Commercial L2, 105 cu ft Royal Mail van at Ards Airport, Belfast, about to be loaded with mail bags which have just arrived from London in a Hillman Airways de Haviland Rapide. This picture was taken c.1937

Nevertheless, Morris dealers were obliged to carry stocks of 'GPO only' parts to cater for the Morris vehicles operated by the GPO in their area and, to add insult to injury, the GPO workshops sometimes 'poached' mechanics from dealers, after they had provided the necessary investment for their training.

Although Morris Royal Mail vans were usually maintained and repaired by the GPO's own workshops, Morris dealers were also called upon to carry out this work, especially in rural areas where the GPO did not have any workshop facilities. Such work was always subject to inspection by GPO engineers before the invoice was accepted and even when accepted, bureaucracy frequently delayed settlement of the account.

The result of all this was that dealers sometimes adopted a less than favourable attitude towards the GPO business, but the large number of Morris and Morris-Commercial Royal Mail vans on the road advertised the durability of Morris products and this fact alone gave dealers the incentive to maintain the business and to accept their difficulties.

A view of the GPO workshop in Manchester. The Morris engine and gearbox mounted on the stand nearest to the camera, is probably an 11.9 h.p. type CB. (see page 40)

13

Car to Commercial Conversions

The popularity of 'Bullnose' and 'Flatnose' Morris tourers during the 1920s prompted several firms to offer convertible commercial bodies for them in order to make them into dual purpose vehicles.

Before a convertible body could be installed, the tourer bodywork had to be modified so that its rear section could be removed completely from the chassis frame. Once this modification had been carried out, however, it took only a few minutes for two people to convert a car into a van or truck.

Some examples of the types of convertible bodies available for Morris cars can be seen in the following advertisements and photographs.

A 1925 'Bullnose' Morris Cowley fitted with a 'Magnet' convertible body

Magnet Patent
Convertible Bodies
for Morris Cars

Use Your Car for Business or Pleasure

Adapting Morris-Cowley Tourer with Magnet Patent and supplying alternate van (both seats detachable) **£56 10 0**

Adapting Morris-Cowley Tourer with Magnet Patent and supplying alternate lorry (both seats detachable) **£33 0 0**

FOR PLEASURE

Adapting Morris-Cowley Tourer with Magnet Patent and supplying alternate semi-van (rear seats only detachable) **£46 15 0**

Adapting Morris-Cowley Tourer with Magnet Patent and supplying alternate semi-lorry (rear seats only detachable) **£22 10 0**

The above prices are for bodies finished Grey priming coat.

In all cases of Magnet conversions to Morris-Cowleys it is necessary to fit an additional door to the rear off-side, but this is a distinct advantage to the rear passengers in getting in or out.

AS A LORRY

Ellison & Smith
LIMITED
MAGNET WORKS
GATLEY, MANCHESTER

WRITE FOR FURTHER PARTICULARS

AS A VAN

MAGNET-IZE   YOUR   MORRIS   CAR

HOW EASILY THE CHANGE IS MADE.

CHANGING a Motor Body conveys to some minds a sense of very disagreeable operations—Looking up tools, unscrewing, heavy lifting, crawling on the floor underneath the car, generally soiling both the clothes and the hands.

But not so with a Magnet Convertible Body. This type of Body stands alone for simplicity, ease of operation and strength.

We will take you simply through the movements.

Imagine for a moment you had returned from a week-end tour in the country late in the evening. The next morning you drive to business, on arriving you desire to immediately proceed to convert your Car into a Lorry for the day's deliveries—with the assistance of a girl you

1. Unscrew two bolts with key as illustrated.

2. Lift out the back seat, placing it on the ground. Lift on the lorry, screw up again and within three minutes you are ready for deliveries.

Could anything be simpler?

SIMPLICITY SIMPLIFIED ——— UTILITY AMPLIFIED

THE "MAGNET" COVERED LORRY.

This model is fitted with detachable tilt cover, with each side and rear curtain made to roll up when not required. Lorry dimensions are 5' 3" long × 4' 11" wide. Loose covers are provided for the front seats. Morris-Cowley adapted and alternative Covered Lorry, supplied painted the same colour as the Tourer .. .. £45 0 0 (at works).

YOU MAY COPE WITH EVERY LOAD IF YOU MAGNET-IZE

THE "MAGNET" LORRY.

The "Magnet" Lorry is light yet very strong, stronger in fact than any other part of the Car. Lorry dimensions are 5' 3" long × 4' 11" wide. Loose covers are provided for the front seats. Morris-Cowley adapted and alternative Lorry complete with Hood, supplied painted the same colour as the Tourer, lined and varnished .. .. £35 0 0 (at works).

SIMPLICITY SIMPLIFIED —— UTILITY AMPLIFIED

❖ 157

MAGNET-IZE YOUR MORRIS CAR

This appears to be an ordinary Morris-Cowley Touring Car. Actually it is fitted with the "Magnet" Patent Convertible Body. Study it carefully and see if you can identify the carefully concealed adaptation.

THEY ARE GOLDEN MINUTES WHEN YOU MAGNETIZE

THE "MAGNET" ICE CREAM FLOAT.

This is constructed of selected hardwood throughout and fitted with a detachable hinged tail board for loading purposes. Loose covers are provided for the front seat. Morris-Cowley adapted and alternative Ice Cream Float, supplied painted in the same colour as Tourer, with lighter panels if desired .. £56 10 0 (at works)
Lettering and decorations are charged for extra according to requirements.

Every Tradesman Owner of a Morris

is certain to be vitally interested in the question of goods transport. Why should his car lie idle in business hours, when quite a small outlay will enable him to purchase our Interchangeable Body—Van, Lorry, Milk Float or Sheep Truck?

We reproduce our "Morris-Cowley" type Van—a most excellent means of publicity for the enterprising tradesman's business. Carrying capacity is 8 cwts. Body changes can be effected, without exertion, in a few minutes, and each type is adaptable to any model of "Morris" car. Guard against disfigurement by purchasing the only "Interchangeable" body.

Bought for their obvious usefulness, distinctive appearance and unique quality, our products carry an unlimited guarantee and can be delivered in 8 hours from receipt of car.

Could better Interchangeable Bodies be built (W&F) would build them

920 ASHTON-U-LYNE
"SERVICE, DUKINFIELD"

Whitehead & Furness Ltd
BODY BUILDERS
MOTOR ENGINEERS
DUKINFIELD

COME INTO THE GARDEN OF ENGLAND
AND EVERY MOTORIST WILL TELL YOU THAT
CHAS. BAKER & CO., of Tonbridge, LTD.
IS THE HOME OF MORRIS SALES AND SERVICE

LARGE STOCKS OF SPARES

ALL MODELS MORRIS CARS AND VANS IN STOCK

MORRIS CARS SALES & SERVICE
CHAS. BAKER & Cº TONBRIDGE LTD
TEL. 105 TONBRIDGE DISTRICT DISTRIBUTORS.

Drive into their Morris Service Works, or 'Phone for their Morris Service Van, fitted with spares, etc., which is ready Day and Night for you at your home :: or on the road ::

Registered Office : 150, HIGH STREET, TONBRIDGE

Telephone 105   Telegrams : Morservice, Tonbridge

ONE OF THE OLDEST MORRIS AGENTS IN THE

Chas Baker & Co
OF TONBRIDGE LTD
PHONE 105
TELEGRAMS: MORSERVICE TONBRIDGE

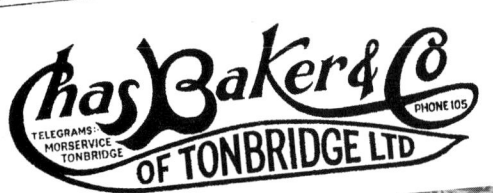

MAIN AGENTS for MID-KENT and Part SUSSEX

One of the oldest Morris Agents in the country ::::::

Our special design of **Convertible Coach built Light Van Body**, made and fitted at our works to standard two and four-seater Cowley cars. Can be altered from Van to Pleasure car in less than ten minutes.

All models of Standard Cowley and Oxford Cars, Light Vans, 12 cwt. and One-ton Commercial Cars in Stock.

Cash or Easy Payments

Registered Office and Showrooms

F. A. SMITH,
GOLD MEDAL BAKERY,
2, BARDEN Rd. & 178, HIGH St.
PHONE 324. TONBRIDGE.

150, HIGH STREET - TONBRIDGE

Jennings SANDBACH
CONVERTIBLE BODIES FOR MORRIS CARS

J. H. JENNINGS & SON LTD.
Motor Body Builders
Sandbach

Phone 62 Established over 150 years
Two-Seater Hood **£5 10s.** extra

Car converted and fitted with flat top - **£15**

Flat top, fitted with sideboards - **£18**

With detachable Sheep Cratches - **£21**

Car converted to Van - **£40**

---

4 - SEATER MORRIS - COWLEY
interchangeable with the
DISTINCTIVE H. W. VAN

Changed in 5 minutes

Complete with Touring and Van body, **£240**
Your present car converted for **£45**

Write for full particulars and measurements— Tel. 35

HOLTBY, WHITE & Co.
Main Morris Agents
7, Prospect Street   BRIDLINGTON   E. Yorks

The Morris 10 cwt 'Flatnose' van chassis (no 382429, erected 4th October 1933) shown on this page and opposite, is one of 164 made during the 1934 season for the GPO (see page 141, note 4) and, in common with all Morris Cowley cars and vans of this period, it is fitted with a 'square'-shaped Calormeter. 'ALX 774' was supplied to the GPO in a batch of seven (ALX 770 to ALX 776) during November 1933, with convertible car/van bodywork manufactured by Duple Bodies and Motors Ltd. and painted in 'Telephone Green' livery. Note that the rear section of the tourer body could be detached, along with the hood assembly, thereby allowing a van body, complete with partition, to be installed. ➙

DUPLE
Convertible
BODIES

(cont.) Before the First World War the Bifort Company, which was owned by Mr Herbert White, began making convertible bodies for cars that they described as dual-purpose and hence coined the word 'Duple'. After the War, White formed 'Duple Bodies and Motors Ltd.' and production of convertible bodies recommenced at a factory in North London. Expansion into other types of bodywork, especially coaches, led to a move to larger premises at Hendon, West London, in 1925. The company continued to manufacture convertible bodies, as well as special bodies for cars, until c.1934, when it ceased these activities to concentrate on making coach bodies

❖ 163

14

Morris Cowley Commercial Travellers Car

The Morris Cowley Commercial Travellers Car, which was introduced during November 1923, was an adaptation of the then current two seater 'Bullnose' Morris Cowley car. A small box van body, in which a travelling salesman could carry a reasonable quantitiy of goods and samples, was constructed behind the seats in place of the dickey.

Similar bodywork was offered on 'Flatnose' Morris Cowleys, from their introduction at the end of 1926, until the body type was replaced by the five-door Morris Cowley Travellers saloon in 1930.

MORRIS COMMERCIAL TRAVELLER'S CAR
Fully equipped with chassis as specified, fitted with Dunlop Cord Balloon tyres and covered by the Morris inclusive insurance scheme (General Accident, Fire and Life Assurance Corporation, Ltd., Policy) - - - Price £190

A 1925 season 'Bullnose' Morris Cowley Commercial Travellers Car

Below: A 1929 'Flatnose' Morris Cowley Commercial Travellers Car

Above & right: The five-door 'Flatnose' Morris Cowley Travellers Saloon, introduced in 1930

❖ 165

15

Conclusion

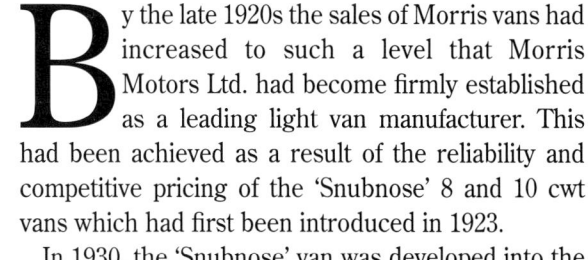

By the late 1920s the sales of Morris vans had increased to such a level that Morris Motors Ltd. had become firmly established as a leading light van manufacturer. This had been achieved as a result of the reliability and competitive pricing of the 'Snubnose' 8 and 10 cwt vans which had first been introduced in 1923.

In 1930, the 'Snubnose' van was developed into the 'Flatnose' and then a new 8/10 cwt van, based on the 1933 season Morris Cowley, was announced for the 1934 season. Up to then, all Morris 8/10 cwt vans had been fitted with similar engines, derived from the Continental 'Red Seal' Type 'U' and 3 speed gearboxes. Then, at the end of 1934, the 8/10 cwt 'TWV' van appeared, which was powered by a different design of engine – the type 'TF' – and, unlike the earlier types with splash fed big end bearings, it had a full force-fed lubrication system and was coupled to a 4 speed gearbox.

The line of 8/10 cwt vans continued with the introduction of the semi-forward control Series 2 'TWV' for the 1936 season, which was developed into the Series 'Y' in 1940. Both the Series 2 'TWV' and the Series 'Y', which remained in production until 1948, had an offset engine and transmission to provide more space for the driver, hence the reason for the offset starting handle position as shown in the illustrations.

Above: Morris Series 2 'TWV' 10 cwt Van, introduced late in 1935

An advertisement illustrating a 'TWV' 8/10 cwt van. This van, which was manufactured for the 1935 season only, appears similar to the 1934 season 8/10 cwt van (see page 94), but it had a different type of engine and gearbox and its bodywork had been redesigned with a new style radiator having the lamp bar passing through the shell. 51 'TWV' Royal Mail vans were operated by the GPO (see pages 141 & 143, note 5)

THOUSANDS
OF MEN LIKE YOU ARE ALREADY RUNNING
MORRIS VANS

...and the 1935 models show still further improvements

- ENTIRELY NEW ENGINES GIVING STILL FINER PERFORMANCES
- INCREASED ACCOMMODATION IN THE 5-CWT. MODEL
- THE LINES AND APPEARANCE OF BOTH MODELS CONSIDERABLY ENHANCED

Speedy, reliable, economical transport—that's one of the trader's biggest problems to-day and fifteen thousand traders have found the answer in a Morris Light Van. The 1935 models give you still more for your money. You can't afford to neglect the extra business and extra prestige the Morris Light Van of to-day will bring you. Send for the descriptive folder and choose the model that's best for your business.

Morris *Light* **VANS**

5-cwt. Model - £115 (Tax £10)
8/10-cwt. Model £165 (Tax £15)
(as illustrated)

IF YOU DON'T BUY MORRIS
—AT LEAST BUY BRITISH

MORRIS MOTORS LIMITED, COWLEY, OXFORD

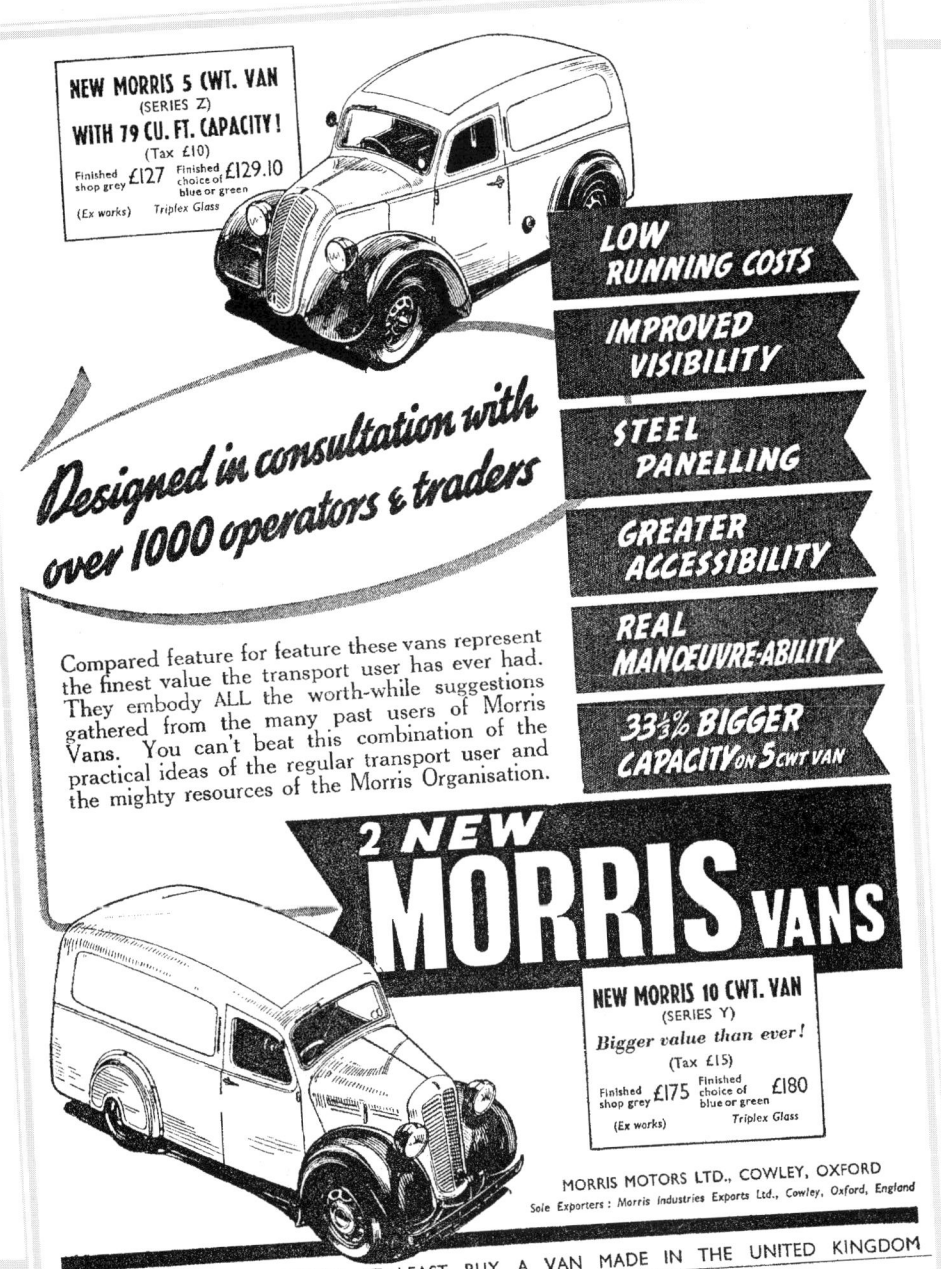

Left: This advertisement appeared in March 1940 and illustrates the Series 'Y' 10 cwt van introduced in that year as well as the Series 'Z', which was derived from the Series E Morris Eight. The GPO operated both Series 'Y' and 'Z' vans

During October 1948, a new 10 cwt van was introduced – the forward control 'J' type – which also featured an offset engine and transmission, but unlike its Cowley-built predecessors, it was made in Birmingham, by Morris Commercial Cars Ltd.

In May 1950, Morris Motors Ltd. introduced the Morris Cowley 10 cwt van, derived from the Series MO Morris Oxford car. This van was restyled and repowered, in 1956, with an overhead valve engine, in place of the previous side valve unit, and designated the Series III, Morris Half-ton Van. Both models were also available as pick-ups.

Left The Morris-Commercial 10 cwt 'J' type van, introduced in October 1948, was also known as the J/L or J/R to differentiate between left and right hand drive models – note the two starting handle holes to suit the offset engine and transmission of each model. When the BMC 'B' series overhead valve engine and 4 speed gearbox replaced the side valve engine and 3 speed gearbox in February 1957 (at chassis no 36266), the vehicle was redesignated the 'JB' and soon after badged as a Morris. In March 1957 (from chassis no 36650) the vehicle was also badged as an Austin and such versions had a redesigned grille panel and were known either a a 'JB/A' or the type '101'. Some vans, designated 'JBO', were built with a BMC 1.5 litre diesel engine primarily for the GPO, who also operated J's and JB's. Production ceased in January 1961 at chassis no. 48621

Right: Introduced in May 1950 and derived from the Series 'MO' Morris Oxford, the Morris Cowley 10 cwt van was the last car-derived light van to be introduced by Morris Motors Ltd. before the formation of the British Motor Corporation

❖ 169

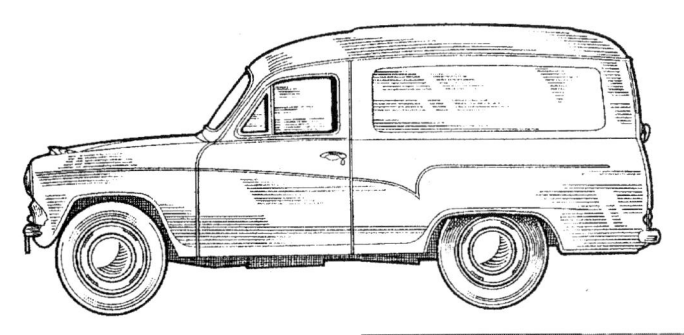

Left: Derived from the Austin A55 car, the Half-ton Van was introduced in 1962 and was badged as either an Austin or Morris. A Half-ton Pick-up was also available

Right: Announced on 1st August 1972 and badged as either an Austin or a Morris to replace the Morris Minor 6/8 cwt and the Austin/Morris Half-ton Van range, the Marina van was offered in two payload capacities – the 7 cwt, having 8" dia. brakes and a semi-floating type of rear axle, and the 10 cwt, utilising a rear axle of three-quarter floating design and 9" brakes. Derived from the Morris Marina car and having a capacity of 88.5 cu ft, both versions were fitted with 1275cc engines, although an engine of 1098cc could be specified for the 7 cwt

This photograph shows two Royal Mail vans; a Morris Marina 2 on the left and a Morris Ital. The Marina 2, 7 & 10 cwt vans, distinguished by their one-piece grille and 'round' headlamps, were introduced in October 1975. During October 1978, these vans were re-designated as the '440' and '575' to indicate their gross payload in kilogrammes, with front disc brakes being fitted from November 1978. In May 1982, the Marina vans were restyled with 'rectangular' headlamps, to bring them up to the specification of the cars from which they were derived, and reintroduced as Morris '440' and '575' Ital vans. The '575' was also available as a pick-up and both the '440' and '575' versions were fitted with 1275cc engines. Production of Ital vans ceased in December 1984, at chassis no. 745418

Following the merger of Morris Motors Ltd. and the Austin Motor Company Co. Ltd. at the beginning of 1952 to form the British Motor Corporation, the Morris Half-ton Van, introduced in 1962 to replace the Series III, was based on the Austin A55 car and badged as both an Austin and a Morris. In January 1968, the British Motor Corporation and the Leyland Motor Corporation announced that they were to merge and create the British Leyland Motor Corporation. Nevertheless, the Cowley factory continued their tradition of manufacturing car-derived vans and in 1972 a range of Austin and Morris Marina 7 and 10 cwt vans were introduced with pick-up versions appearing two years later. From 1980, Marina vans displayed British Leyland trade marks and then, in 1982, they were restyled with 'rectangular' headlamps and re-introduced as Morris '440' and '575' Itals. When production of the Ital van ceased in December 1984, 70 years after the introduction of the first Morris car-derived van, the Morris marque became extinct.

Index

Illustrations are indicated by italic numerals

A.C. Cars 98
Adderley Park 4, 57, 112, 113
Admiralty 15, 24, 35
Ainsworth, H. M. 36
Allen, Alfred 29
Angus Sanderson & Co., Sir William 27, 96, 97
Ards Airport 152
Armstead, Edward 20, 98
Austin Motor Co. Ltd. 25, 27, 33, 37, *147*, 171
Austin A55 .... 170, 171
Autocar, The 21
Axle 19, 21, 25, 34, 59, 70, 74, 76, 80, 84, 88, 92, 96, *97*, 100, *104,* 109, *119,* 128, 131, *138, 139, 140, 144,* 146

Baico 131, *131*
Bainton Road, Oxford 44
Bangor 132
Barton, F. G. 20, *20*
Barton Motor Co. Ltd. 20
Bedford Head Postmaster 124
Benet, M. 36
Birmingham 4, 21, 33, 49, 57, 96, 103, 112, 115, 117, *147,* 168
Blake, Edgar H. 29, *29,* 99
Boden, Oliver 100, 115, *115*
Bodies *13,* 21, 25, 31, 50–2, *53,* 59, 63, 70, 72, 74, 76, 78, 80, 84, 88, 92, *107, 108, 109,* 110, *111,* 112, 114, *118,* 120, 122, 123, *125, 128, 129, 130,* 131, 132, *133, 134,* 138, 139, *142,* 143, 144, *145, 149, 151, 154–63,* 164, 167
convertible commercial
  Chas. Baker & Co. 160
  Ellison & Smith Ltd. (Magnet) 154–8
  Duple Bodies and Motors Ltd 162–163

Holtby, White & Co. 161
J. H. Jennings & Co. 161
Whitehead & Furness Ltd. 159
Brakes 17, 30, 64, 70, 74, 76, 80, 84, 88, 92, 94, 110, *117, 118, 121,* 128, 130, *137,* 140, 144, 146
British Industries Fair 117
British Leyland Motor Corporation ... 171
British Motor Corporation 33, 37, *169,* 171
British Overseas Airways Corporation 115
British Spyker Co., Ltd. 97
Burgess, W. H. M. 26, 27, *57,* 58

Carburetters
  S.U. '2M' *54,* 56, 57, 58, 74, 76
  'HV2' 58, 76, 80, 84, 88, 92
  'Sloper' 55, *55,* 56, *56*
  Smiths '26HKMC' 58, 64, 144, 146
  '4MO' 58, 64
  'Straight-through' 58, 64, 70, 74
  Solex 58, 144, 146
Carter Paterson *145*
Cannell, William 100, *100*
Charlesworth Motor Bodies 98
Chassis 11, 12, 14, 17, 52, 59, *83,* 98, 100, 110, *117,* 120, 122, 123, 126, 131, 132, 138, 139, 140, 141, 143, 144, 154
frame 63, 64, 65, 70, 74, 76, 80, 84, 88, 92
identification 60
GWK 127
'T' type one ton *100,* 108
Chevrolet 14
Clinch, M. F. *51*
Clutch plates 88, 92, 146
Cooper, Joseph 19
Commercial Motor, The 10, 11, 12
Continental Motor Manufacturing Co. . 34
Coventry-Simplex 126

Cowley factory 23–33, *27, 41, 51, 52, 53, 102*
Cranham, H. W. 26
Creke, W. Launcelot 19
Croydon Aerodrome *118, 150, 151*

Davis, A. L. 42
Dagenham 131
Detroit Gear and Machine Co. 35, 36
Drymen *137*
Dodge 14
Doherty Motor Components Ltd. 42
Dunlop Rubber Co., Ltd. 99
Dynamo 123, 132, 141
Dynamotor 123, 132, 141

Eastbourne 116
Engine 4, 14, 16, 19, 26, 30, 36, 38, *41,* 54, 55, 56, 59, 96, 126, 128, 140, 148, *153*
Continental 4, 13, 24, 26, 34, *35,* 36, 50, 148
cylinder head ('turbulent') 40, 80, 84, 88, 92, 140
  'non-turbulent' 40, *148*
identification 40
mountings 76
type 'CA' 13, 36, *37,* 40
type 'CB' 40, 63, 64, 70, 74, 76, 109, 132, 139, 140, 146, 148, *153*
type 'CD' 40
type 'CE' 40, 100, *100,* 110
type 'CG' 40, 80, 84, 140, 148
type 'CH' & 'CL' 40
type 'CJ' & 'CM' 40
type 'CK' 40, 143, 148
type 'CN' 40, 88
type 'CO' 40
type 'CP' 40, 143, 146, 148
type 'CQ' 40, 143, 146, *148*

type 'CR' 40
type 'CS' 40, 92, 143
type 'CSDC' 40, 148
type 'IM' & 'MM' 40
type 'TF' 143
1098cc 170
1275cc .170, 171
White & Poppe 4, *6*, 13, *13*, 21, *22, 23*, 26, 34, 50, 96, 126

Farman Automobile Co. 19
Ford 4, 14, 15, 16, 29, 30, 116, 119, 122, 123, 130, 131, 149, 152
   Eight 29, 30
   Model 'A', 'AF' & 'AA' 4, 16, 123, 130, 131
   Model 'T' & 'TT' 4, 10, 14, *14*, 16, 17, *17*, 97, 116, *119*, 120, 122, 123, 130, *130*, 131, *131*, 149
   7 cwt Delivery Van 14, *16*, 17, 131
   One Ton Truck 15, *17*, 109
Forth Railway Bridge 135
Frederick, W. A. 34, *34*
Fuel Tax 16
Fulford, William 50
Fume consumer 92

Gearbox 21, 25, 34, 36, *37*, 38, 59, 64, 70, 74, 76, 80, 84, 88, *90*, 92, 96, 100, 110, 128, 130, 139, 143, 146, 166
George Street, Oxford 20
George Wailes & Co. 54
General Strike, The 33
Gosford Street 36–9, *36, 37, 39*
Grey, H. W. 99
Grice, Arthur 126
GPO 4, 63, 88, 100, *101*, 110, *110*, 116–53, 169
GWK 116, 122, 123, 126-7, *127*

Hamill, William Wilson 100
Haslemere Sorting Office 149
de Havilland Rapide 118, 152
Hewen's Garage 12
Hobbs, Wilfred 99
Hollick & Pratt Ltd. 25, 31, 50–52, *51*, 114
Hollick, Edward 50
Hollow Way, Cowley 102
Hotchkiss et Cie 4, 25, 31, 34, 36, 38

Import Duties 4, 14, 15, 16, 34
Institute of Automobile Engineers 98

Kendrick, J. A. 96
Keiller, C. M. 126
Kimber, Cecil 20, 32, 44, 98, *98*
King George VI 22
Kingston-upon-Thames 116, 128
Kings Cross Station 98
Kopalapso Folding Roof 62

Lamps 63, 65, *66*, 83, 88, 92, 94, *116, 117, 134, 139, 141*, 144, *149, 167*
Land's End Hotel, The 121
Landstad, Hans 24, *24*, 34
Lawrence-King, C. F. 100, *100*
Leyland Motors Ltd. 128
Leyland Motor Corporation 171
Lilley & Skinner 54, 56
Liverpool 118
Loch Lomond 137
Longwall 20, *20*
Lord, Leonard P. 29–33, *28, 29*, 37, 115

Macclesfield, Earl of 26
Manchester *153*
   Technical School 98

Martinsyde Aircraft 98
'Maxfield' tyre pump 40
McKenna, Reginald 34
Maudslays 123
Mine sinkers 24, *24*, 25
Ministry of Munitions 7, 24
M.G. Car Co. Ltd. 32, 44, 98
M.G. 14/28 44
M.G. 14/40 40
M.G. 'M' type Midget 97, 100, 108
Morris Commercial Cars Ltd 15, 16, 21, 30, 32, 44, 48, 58, 96–113, *107, 109, 111*, 115
Morris-Commercial
   Imperial *113*
   'L' Type 12 cwt 4, 40, *106*, 109, 110, 122, 123, *134, 135*, 138, 139, 143, 145, *146*
   L2 . . . . . . 4, 40, 109, 110, *117*, 122, 123, 138, *139, 142*, 144, *145*
   L2/8 . . . . 4, 40, 110, 123, 138, *142*, 143, 144, 145
   L3 . . . . 40, 110, *110*, 113, 145
   L3/8 . . . . 110, 145
   LT 40, 110
   'T' type tonner 4, 15, 40, 100, *102, 105, 107*, 108, *109*, 110, *110, 111, 112*, 123
   T2 . . . . 4, 40, 110
   T3 . . . . 40, 110, *113*
Morris
   'Bullnose' 3, 4, 28, *43*, 60, 70, 98, 132, 154
   'Bullnose' Vans 13, *35*
   Commercial Travellers Car . . . 164, 165
   Commercial Travellers Saloon
      164–165
   Cowley 4, 13, *24*, 25, 29, 33, 34, 40, 46, 47, 50, 52, 58, 64, 70, 74, 76, 80, 83, 88, 92, 132, 140, 143, 164

Cowley 'Bullnose' 4, 12, 28, 33, 44, 46, 52, *61,* 63, 64, 109, 132, 138, *154,* 164
Cowley 'Flatnose' 4, 40, *47,* 52, 58, *62,* 63, 74, 76, 80, 82, 98, 138, 140, 141
Cowley 10 cwt (1950) 168
Eight *29, 30, 31, 168*
'Flatnose' 3, 33, 44, 47, 60, 63, 70, 121, 123, 139, 140, 141, 144, 154,
'Flatnose' Light Van 4, 40, 80, *81, 82, 83,* 84, *85, 86, 87, 88, 89, 90, 91,* 123, 139, *136, 137,* 138, 140, 141, 144
Half-ton Van 170, 171
Ital Van 171, *171*
'J' Type Van 168, *169*
Marina Van *170,* 171, *171*
Minor 30, 100, *101,* 122, 123, *124, 125*
Minor 6/8 cwt Van 170
MO Oxford 168, *169*
Oxford 4, 13, 21, *23,* 26, 28, 29, 44, 46, 56, 57, 64, 96, 100, 140
Oxford 'Bullnose' 12, 28, 33, 44, 46, 100
Oxford 'Flatnose' 4, 40, 98, 140
Series 'Y' *166,* 168
Series 'Z' *168*
Series III Morris Oxford 168
'Snubnose' 3, 4, *8,* 10, *10,* 13, 17, 23, 35, 40, 44, *45,* 46, 48, 58, 60, 61, 62, 63, 64, 65, *66, 67, 68, 69,* 70, 71, *72, 73,* 74, *75,* 76, *77, 78, 79,* 102, 112, *117, 118,* 119, *120,* 123, 132, *133, 134,* 138, 140, 148, *149,* 166
'TWV' 92, 143, 148, 166, *166, 167*
440 Van 171
575 Van 171
8/10 cwt Van 4, 63, 92, *93, 94, 95,* 166

Morris, Frederick 19
Morris Garage, The 20
Morris Garages, The 14, 21, *21,* 44, 51, 98
Morris Industries Ltd. 31, 32, 58
Morris Industries Exports Ltd. 32, 33
Morris Motor Cycle 19, *19,* 20
Morris Motors Ltd. 4, 8, 10, 13–16, 19, 26–31, *27,* 36, 37, 38, *41,* 42, 44, 50, 51, 52, 58, 59, 99, 100, 103, 112, 114, 115, 116, 117, 120, 123, 132, 140, 143, 152, 166, 168, 169, 171
Radiators Branch 31, 44
Engines Branch 31, *36,* 38, *39*
Bodies Branch 31, 50, 52
Morris Motors (1926) Ltd. 25, 31, 32, 44
Morris Owner, The 8, 57, 115, 117
Morris, Sir William 3, 18, 28
Morris, William Richard 3
Morris, W. R. *3,* 8, 14, 15, 18–33, *18, 19, 28, 29,* 34, 36, 38, 42, 44, 50, 52, 56–8, 96–115
Motor, The 28
Motor Transport 14, 103

Nuffield Exports Ltd. 32
Nuffield, Lord 3, 18
Nuffield Organisation 32, 115

Odometer 117, 139, *139,* 151
Olympia Motor Transport Show 13
Osberton Radiators Ltd. 25, 31, 42–4, *43*
Ostwalt, G. E. 97
Overland 14, *16*
Oxford Automobile and Cycle Agency, The 19
Oxford Garage, The 20
Oxford Motor Omnibus Company 42

Oxford University Flying Squadron ... 42
Parker 18
Paskell, E. C. 116, *116*
Peterborough *118*
Pistons
cast iron 140
aluminium 140
Post Office Telephones 101, 123, 128, 132, *145, 162, 163*
Pratt, L. W. 50–2, 99, 114, *114*

Quarterman, Frank *51*
Queens Ferry 135
Queen Street 21, *21*

Radiators 42–4, *43,* 64, 70, 74, 76, 80, 84, 88, 92, 100
8/10 cwt 47
badges 48, 110, *112*
drain taps 141
'Flatnose' 47
'Snubnose' 46
temperature gauges *49,* 162
'T' type one ton truck *112*
Raworths of Oxford 21, 50, 51
Ricardo 40, 80, 84, 88, 92, 140
Rose, P. G. 100, *101,* 103
Rowse, Arthur 24, 25, *25*
Royal Mail 4, *101,* 110, 116–53, *171*
Ryder, H. A. 42, *43,* 44

Sanction number 144
Sheffield Simplex 98
Shell 61
Shock absorbers
Armstrong 92
Gabriel Snubbers 123
Skinner, George Herbert 54, 56
Skinner, Thomas Carlyle 54–8, *54*
Southport Grammar School 98

Spare wheel 138–39
Specialisation 30, *31*
Specialloid Pistons Co. 98
Speedwell Co. 19
Springs, Road 64, 70, 74, 76, 80, 84, 88, 92, 100, 128, 132, 140
Star 54
Steering . . . . 21, 34, 96, 100, 128, 144, box 64, 70, 74, 76, 80, 84, 88, 92, 146
Stevens, J. D. 96
Stewart & Ardern *22,* 23, 26
Stewart, Gordon *22,* 23, 26, *102*
Stokes bombs 24
Studebaker 14
S.U. *54, 55, 56, 57,* 74, 76, 80, 84, 88, 92
S.U. Company, Ltd. 26, 31, 54–58
   Petrolift 92
   petrol pump 92
Super Tax 30, 31, 32

Talfourd Wood, J. 126
Thomas, W. M. W. (Miles) 32, 100, 115, *115,*
Thornycroft J type 7
Thornton, Reginald W. 32, *32,* 99
Trade marks 112
Trafficators *140*
Trafford Park 14, 17, 64, 130, 131
Trojan 117, 122, 123, 128, *128,* 129, *129*
Tylor & Sons Ltd., Messrs. J. 96
Tyres 64, 65, 146, *164*
   beaded edge 65, 132
   Dunlop cord 108, *108*
   pneumatic 103, 108, *108*
   solid 108, *108,* 128

Unipart Group 44

Upjohn, George H. 22

Varney, Mr. 22

Wall Street, financial crash 29
Walsh, Andrew 32, *32,* 98
Weston-Super-Mare 149
W. Harold Percy, Ltd. 101
Wheelbase 64, 70, 74, 76, 80, 84, 88
White & Poppe 4, 13, *13,* 21, 22, 23, *23,* 26, 34, 50, 96, 126
Windscreen 64, 66, 70, 74, 76, 80, 84, 88, 90, 92, 139
Wings 63, 64, 70, *70,* 74, 76, *76,* 80, 84, 88, 92

Wolseley Aero Engines Ltd. 32, 100
Wolseley Motors Ltd. 4, 27, 29, 32, 33, 44, 55, 56, 57, 97, 112, 115
Wood, Paul 129
Wood, Reginald 129
Woodstock Road, Oxford 44
Woollard F. G. 21, 38, *38,* 96
Wrigley & Co., Ltd., E. G. 4, 21, 25, 27, 38, 96–112, *97*
Wrigley, Edward Greenwood 96
W. R. M. Motors Ltd. 4, 13, 19, 23, 26, 58

List of Tables

Commercial vehicles in use in the UK	9
Morris 8 cwt Light Van & Ford Model 'T' 7 cwt Delivery Van production	17
Percentage of total UK car sales	30
Pre-tax profits of the businesses controlled by W. R. Morris	33
Hotchkiss/Morris 11.9/13.9 h.p. engine types	40
Engine specifications	40
Deliveries of vans to the Royal Mail Fleet – 1920 to 1936	122
Morris & Morris-Commercial 70 & 105 cu ft Royal Mail Van chassis supplied between 1924 to 1935	138
Specification for Morris-Commercial 'L' 12 cwt, 'L2/8' & 'L2' 15 cwt chassis	146
Hotchkiss/Morris 11.9/13.9 engine types used in 70 & 105 cu ft Morris Royal Mail Vans – 1924 to 1940	148

Acknowledgements

The author is most grateful to the following for providing information and/or illustrations for this book:-

Robin Barraclough
Barry Blight
John Breach
David Burgess-Wise
Graham Bushnell
Carl Cederstrand
Tony Cripps
Harry Edwards
Geoffrey Fishwick
Philip Garnons-Williams

John Leach
Michael Lowndes
Charles Moody
Brian Moore
Norman Painting
Walter Peters
Bernard White
Jonathan Wood
Paul Wood

The British Motor Industry Heritage Trust
The Bullnose Morris Club
Burlen Fuel Systems Ltd.
The Commercial Motor
Glasses Information Services Ltd.
The Morris-Commercial Club
The Morris Register
Motor Transport
The Post Office
The Post Office Vehicle Club
Ricardo Consulting Engineers Ltd.
Vicarys of Battle Ltd.

Published in 1999 by P&B Publishing,
32 High Street, Battle, East Sussex, TN33 0EH

Copyright © P&B Publishing

ISBN 0 9536327 0 9

Designed and typeset by Brian Folkard Design,
10 Maryland Way, Sunbury-on-Thames,
Middlesex, TW16 6HR

Printed in Great Britain by Jaggerprint/Victoria,
Kingston-upon-Thames, Surrey KT2 5EW

All rights reserved. No part of this publication may be reproduced, stored in a retrieval system, or transmitted in any form or by any means; electronic, mechanical, photocopying, recording, or otherwise; without the permission of the publishers